Nonlinear Waves
and
Inverse Scattering Transform

Nonlinear Waves
and
Inverse Scattering
Transform

Spencer Kuo
New York University, USA

 **World Scientific**

NEW JERSEY · LONDON · SINGAPORE · BEIJING · SHANGHAI · HONG KONG · TAIPEI · CHENNAI · TOKYO

Published by

World Scientific Publishing Europe Ltd.

57 Shelton Street, Covent Garden, London WC2H 9HE

Head office: 5 Toh Tuck Link, Singapore 596224

USA office: 27 Warren Street, Suite 401-402, Hackensack, NJ 07601

Library of Congress Cataloging-in-Publication Data

Names: Kuo, Spencer P., author.
Title: Nonlinear waves and inverse scattering transform / Spencer Kuo.
Description: New Jersey : World Scientific, [2023] | Includes bibliographical references.
Identifiers: LCCN 2023000255 | ISBN 9781800614031 (hardcover) |
    ISBN 9781800614048 (ebook for institutions) | ISBN 9781800614055 (ebook for individuals)
Subjects: LCSH: Nonlinear waves. | Inverse scattering transform.
Classification: LCC QA927 .K863 2023 | DDC 531/.1133--dc23/eng20230508
LC record available at https://lccn.loc.gov/2023000255

British Library Cataloguing-in-Publication Data
A catalogue record for this book is available from the British Library.

For any available supplementary material, please visit
https://www.worldscientific.com/worldscibooks/10.1142/Q0413#t=suppl

Desk Editors: Logeshwaran Arumugam/Adam Binnie/Shi Ying Koe

Typeset by Stallion Press
Email: enquiries@stallionpress.com

Printed in Singapore

To my parents,
Wen-Hsiu and U-Lan Kuo
and to Sophia Ping Kuo,
my wife, constant companion, and best friend.
She is the loving mother to my daughter Adeline,
grandmother to Wesley Wong,
and mother-in-law to WaiKin Wong.

Preface

Nonlinear waves are essential phenomena in scientific and engineering disciplines. The features of the nonlinear waves are usually described by solutions to nonlinear partial differential equations, which are fundamental to students and researchers.

This book was prepared to familiarize students with nonlinear waves and methods of solving nonlinear partial differential equations, which will enable them to expand their studies into related areas. The selection of topics and the focus given to each provide essential materials for a lecturer to cover the bases in a nonlinear wave course.

The first chapter introduces different "mode" types in nonlinear systems in terms of the nonlinear waveforms as well as the generic nonlinear partial differential equations (NLPDEs). Bäcklund transform is introduced to solve different types of NLPDEs for stationary solutions. The ways of finding Bäcklund transforms are illustrated. Chapters 2 and 3 are devoted to the nonlinear systems descriptive by three generic nonlinear equations: nonlinear Schrödinger equation, Korteweg–de Vries (KdV) equation, and Burgers equation, which are derived and analyzed analytically. Characteristic features of the solutions of the nonlinear equations are presented. Methods to obtain periodic and solitary solutions are introduced. The non-stationary solitary solution of the three-dimensional nonlinear Schrödinger equation is discussed. Chapter 4 is devoted to the inverse scattering transform (IST), which is a powerful technique to address the initial value problems of a group of nonlinear PDEs.

The concept and formulas of IST are presented; Lax equation and AKNS equation, which represent an NLPDE, subject to a specific pair of operators (Lax pair and AKNS pair, respectively), are introduced. KdV equation is adopted to illustrate methods to determine the pair of operators, which lead to two auxiliary equations: one is in the form of a linear Schrödinger equation to define the scattering by the solution function of the NLPDE; the second one is a rate equation, which updates the scattering data of the linear Schrödinger system (the first auxiliary equation) having the initial condition of the NLPDE as the scatterer. IST then reconstructs the potential function (scatterer) of the Schrödinger equation via the scattering data. In Chapter 5, derivations and proof of the formulas used in IST are presented. Steps for applying IST to solve NLPDE for solitary solutions are illustrated in Chapter 6.

I wish to express my sincere gratitude to Professors Bernard R. S. Cheo and Nathan Marcuvitz who guided me in this field and have given me much helpful advice and kind encouragement on my scientific evolution throughout my career. Some of the examples presented in this book are references to the class notes prepared by Prof. Marcuvitz for a "nonlinear waves" course. I would like to acknowledge Dr. Arnold Lee Snyder for collaborating on the observation of upper hybrid caviton in ionospheric heating experiments at HAARP, which is presented in Chapter 2, and Professor Min-Chang Lee for lasting research collaboration.

About the Author

 **Spencer Szu-Ping Kuo** received B.S./M.S. degrees from National Chiao Tung University, Taiwan, ROC, in 1970/1973, and Ph.D. degree in 1977 from Polytechnic Institute of New York (now known as New York University-Tandon School of Engineering). Since 1986, he has been a Full Professor in ECE Department, Polytechnic University (became NYU-Tandon), until he retired in 2020 as a Professor Emeritus. He initiated and directed a "summer research program for college juniors" in the ECE department from 1985 to 1991. This program has gained nationwide popularity and has been adopted in many universities and national laboratories. His research contributions include a demonstration of photon acceleration by a rapidly created plasma and the creation of plasma crystals to trap photons. He invented various plasma torches for igniting supersonic combustors and mitigating sonic booms of supersonic-aircraft; he invented an air plasma spray for wound-bleeding control and wound healing, and as a disinfectant. Dr. Kuo is a fellow of the IEEE. He was Principal Investigator for more than 30 research projects awarded by the NSF, AFOSR, ONR, and NASA. He published four textbooks and more than 210 journal articles and holds 11 patents. He received an outstanding research award from Sigma Xi in 1990. The Chinese Institute of Engineers named him a 2005 Asian-American Engineer of the year. He was honored as a "distinguished alumnus" of his Alma Mater, National Chiao Tung University in 2013.

Contents

List of Figures

Chapter 1

Nonlinear Waves

1.1 Introduction

Waves are the coherent progressive variation of physical quantities in space and time. Water waves are the most readily perceived example as appearing in the form of a space-time-dependent disturbance moving on the water surface; on the other hand, light and sound are the two most important waves for sensing the world around us.

The physical quantities appearing in each wave system depend strongly on the properties of the media supporting their propagation of which the combined effect of the nonlinearity and dispersion on the wave propagation is considered in this book. The effect of dissipation is included in a few cases.

Nonlinear wave phenomena occur in the propagation of waves of high and low frequencies through media, where the wave-dependent changes in the properties of media become significant. The solutions of the nonlinear wave equations often exhibit unique phenomena, such as stable localized waves (e.g., solitons), self-similar structures, chaotic dynamics, and wave discontinuities such as shock waves (derivative singularities), and/or wave collapse (where the solution tends to infinity in finite time or finite propagation distance). Nonlinear waves are of wide physical and mathematical interest and practical applications in a variety of areas, such as nonlinear optics, fluid dynamics, and plasma physics.

1

A distinction between linear systems and nonlinear systems is that the superposition principle is not applicable for nonlinear systems. A wave perturbation in a linear system can be obtained via a linear combination of the eigenwave modes, which follow the linear wave dispersion relations of the system and do not interact with each other. On the other hand, it is not possible to predict the evolution of the nonlinear wave based on the characteristics of these linear eigenmodes because these modes interact with one another. Moreover, nonlinear media do not, in general, admit constant speed propagation of waves with arbitrary amplitude and shape. There exist different categories of nonlinear media; for certain amplitudes, some admit the propagation of (periodic or pulse) waves of definite shape with a constant speed; in others, the admitted waves have neither a definite shape nor a constant speed. Waves that can be propagated with constant speed and shape are called stationary waves of which those in the forms of solitary and nonlinear periodic are of interest in the study of nonlinear waves and their applications. On the other hand, nonlinear waves have neither a constant speed nor shape, that is, they are "non-stationary".

When a wave is impinging on a discontinuity (spatial or temporal) in a linear medium, it gives rise to a multiplicity of scattered waves. The scattered waves are characteristic of the various "channels" associated with the discontinuity. The determination of the scattered waves, their amplitudes, forms, etc., is well developed for linear systems.

In nonlinear systems, there occur, in addition to space-time discontinuities, "self-induced discontinuity" regions wherein the amplitude or phase of the wave field changes very rapidly, or where the amplitude suddenly becomes very large. In either event, there arises a multiplicity of scattered waves. One must distinguish, as in the linear case, between the waves or "mode types" that can exist in the various "channels" associated with the discontinuity or transition region. For example, an incident wave on such a discontinuity may be a non-stationary nonlinear wave that steepens and breaks at the discontinuity, then on leaving the transition region, it decomposes into different types of stationary waves. The determination of the wave behavior in the transition region at the discontinuity is a major challenge. A general wave description near a discontinuity is complicated by the presence of several wave types, i.e., the differential equations are of high order. The reduction of these equations to a pair of generic first-order equations, via Bäcklund transform as presented in Section 1.3, descriptive of different wave types, is a necessary first step. These lead to the methods of characteristic curves and Riemann invariants in some cases.

Waves in nonlinear media may be of low frequency or high frequency. A nonlinear partial differential equation (NLPDE), which models a nonlinear system, may be a baseband equation for low-frequency wave packets or an envelope equation for high-frequency wave packets, where envelope represents the modulation amplitude of the carrier of a high-frequency wave packet.

The determination of the possible stationary waves, and more generally of nonstationary waves, in each nonlinear medium is a basic kinematic concern. The analytical technique for their determination is dependent upon the nature of these waves, for example, whether they are periodic, aperiodic, or quasi-periodic. Periodic or aperiodic waves of stationary type may be sought by solving the space-time (x, t)-dependent partial differential wave equation in a constantly moving wave frame, defined by a coordinate $\xi = x - \mathcal{U}t$ or a dimensionless one $\xi_1 = kx - \omega t$, traveling with the wave (traveling wave approach). In terms of ξ or ξ_1, the wave differential equations are ordinary, at least in one dimension, and may be solved by quadrature in many cases.

Stationary solutions, if they exist, constitute the "mode" types of a nonlinear system. In a suitably moving frame, all are wave solutions that have an invariant form, namely $f(x, t) = f(\xi)$. Stationary solutions may be periodic or aperiodic. In the linear limit when the wave amplitude approaches zero, periodic waves become sinusoidal; solitons and shocks no longer are present (i.e., they degenerate to a trivial "constant"-type solution). Note that the wave structure (amplitude, velocity, etc.) is constant in space and time for a stationary solution.

Examples of both types (periodic and aperiodic for baseband and envelope) of stationary nonlinear waves are displayed in the following:

(1) For low frequency or baseband wave phenomena in nonlinear systems, stationary solutions of NLPDEs take the forms shown in Figs. 1.1(a)–1.1(e).

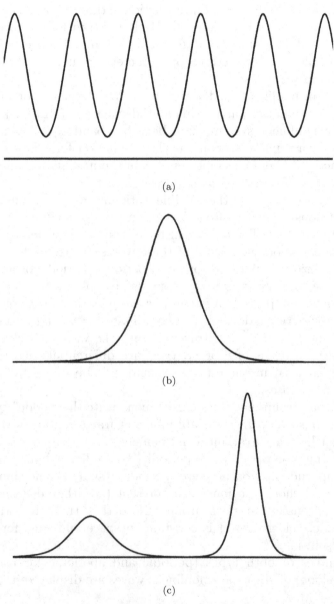

Fig. 1.1. Periodic and aperiodic wave solutions for nonlinear systems. (a) Periodic, non-sinusoidal (Jacobi elliptic function-many sinusoidal harmonics). (b) Single soliton. (c) Two solitons. (d) Shock ("dissipation" necessary to change state and form shock). (e) Oscillatory (dispersive) shocks with leading/trailing wave train.

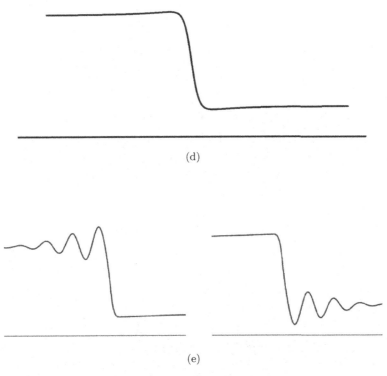

(d)

(e)

Fig. 1.1. (*Continued*)

(2) For high-frequency or carrier wave phenomena in nonlinear systems, stationary solutions, including carriers, take the forms shown in Figs. 1.2(a)–1.2(c).

In the presence of a source of excitation, the wave structure of the basic stationary solutions will be perturbed and become a weak space-time variable. These wave-packet (or local stationary wave) solutions may arise in certain regions of space-time or at transition regions characterized by a rapid change of the amplitude and/or its derivatives. *Such regions do not arise in linear, homogeneous, stationary systems.* On the other hand, a rapid change of the wavelengths and/or frequencies does occur in the presence of spatial or temporal discontinuities. Thus, inhomogeneities or discontinuities in linear systems create effects somewhat like amplitude discontinuities in nonlinear systems.

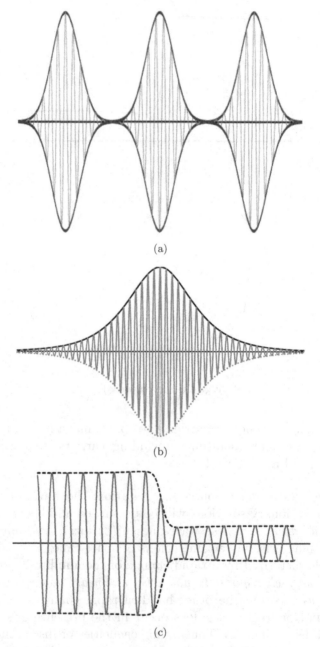

Fig. 1.2. High-frequency nonlinear wave phenomena. (a) Periodic modulation. (b) Soliton modulation. (c) Shock modulation.

1.2 "Mode" Types in Nonlinear Systems (Riemann Invariants)

In actual physical problems, if stationary solutions exist, fields are composed of many "mode" types; each type is characterized by its own speed and amplitude structure. They may or may not be independent of one another and indeed are coupled in "collision" regions wherein changes in parameters occur. The general problem is very complicated, and hence, one first tries to understand generic single "mode" problems described by typical field equations of the following form:

1. Nonlinear and non-dispersive

$$\frac{\partial \phi}{\partial t} + \phi \frac{\partial \phi}{\partial x} = 0$$

2. Nonlinear and dispersive (Korteweg–de Vries (KdV) equation)

$$\frac{\partial \phi}{\partial t} + \phi \frac{\partial \phi}{\partial x} + \alpha \frac{\partial^3 \phi}{\partial x^3} = 0$$

3. Nonlinear with damping but no dispersion

$$\frac{\partial \phi}{\partial t} + \phi \frac{\partial \phi}{\partial x} + \alpha |\phi|^2 = 0$$

4. Nonlinear and dispersive with damping (Burgers equation)

$$\frac{\partial \phi}{\partial t} + \phi \frac{\partial \phi}{\partial x} + \alpha \frac{\partial^2 \phi}{\partial x^2} = 0$$

5. Nonlinear and dispersive without damping (nonlinear Schrödinger equation)

$$i \left(\frac{\partial \phi}{\partial t} + v \frac{\partial \phi}{\partial x} \right) + \alpha \frac{\partial^2 \phi}{\partial x^2} + \beta |\phi|^2 \phi = 0$$

Nonlinearities usually occur at high power levels so that the material media supporting the waves are modified considerably by the propagating waves; consequently, the wave propagation and its characteristics undergo self-modification. Hence, the analysis requires composite, self-consistent descriptions of both wave and the material systems.

1.3   Analytical Solutions of Nonlinear Wave Equations via Bäcklund Transform (Stationary Solutions)

A Bäcklund transform is typically a system of first-order partial differential equations relating to two functions and often depending on an additional parameter. Each of the two functions is a Bäcklund transformation of the other; in general, the two functions separately satisfy partial differential equations (PDEs).

On the other hand, a Bäcklund transform which relates solutions of the *same* equation is called an invariant Bäcklund transform or auto-Bäcklund transform. Such a transform leaves a partial differential equation invariant. It provides a method to transform a known solution of the partial differential equation to a different second solution of the same equation. Hence, much can be deduced about the solutions of the equation, especially if the transform contains a parameter.

Bäcklund transform reduces a high-order partial differential equation to a system of two first-order (of x and t derivative, respectively) partial differential equations, which become solvable analytically. It is a powerful method for finding solutions to NLPDEs. However, no systematic way of finding Bäcklund transforms is known.

In the following, examples are presented to illustrate ways of finding Bäcklund transforms for different PDEs and applying those to obtain solutions to corresponding PDEs.

1.3.1   *Korteweg–de Vries (KdV) equation*

$$\frac{\partial}{\partial t}\phi + \frac{\partial^3}{\partial x^3}\phi + 6\phi\frac{\partial}{\partial x}\phi = 0 \qquad (1.1)$$

The KdV equation was first formulated to describe shallow water wave propagation. It is a fundamental mathematical model for the description of weakly nonlinear long wave propagation in dispersive media; it possesses exact analytic solutions of which solitons and nonlinear periodic traveling waves are of significance in the study of long internal waves in a density-stratified ocean, ion acoustic waves in a plasma, and acoustic waves on a crystal lattice.

Bäcklund transform is derived in the following and applied to find soliton solutions to the KdV equation.

A. Derivation of Bäcklund transform of direct form

A function u with the property $u_x = \phi$ is introduced, where $u_x = \frac{\partial}{\partial x}u$. This function is then substituted into (1.1) to convert the KdV equation to the "potential KdV" equation:

$$\frac{\partial}{\partial t}u + \frac{\partial^3}{\partial x^3}u + 3u_x^2 = 0 \tag{1.2a}$$

which is rearranged as

$$\frac{\partial}{\partial t}u + \frac{\partial^3}{\partial x^3}u + 3u_x^2$$

$$= u_t + 2u_x^2 - uu_{xx} + uu_{xx} + u_{xxx} + u_x^2$$

$$= (u_t + 2u_x^2 - uu_{xx}) + \frac{\partial}{\partial x}(uu_x + u_{xx})$$

$$= (u_t + 2u_x^2 - uu_{xx}) + \frac{\partial^2}{\partial x^2}\left(\frac{1}{2}u^2 + u_x\right) = 0 \tag{1.2b}$$

A Bäcklund transform is introduced as follows:

$$v_x = u_x + \frac{1}{2}u^2 \tag{1.3a}$$

$$v_t = u_t + 2u_x^2 - uu_{xx} \tag{1.3b}$$

and (1.2b) becomes

$$\frac{\partial}{\partial t}v + \frac{\partial^3}{\partial x^3}v = 0 \tag{1.3c}$$

Exercise 1.1: Find a general traveling wave solution of (1.3c).

Ans: $v(x,t) = A\cos\alpha(x + \alpha^2 t) + B\sin\alpha(x + \alpha^2 t)$, where A and B are constant parameters and α^2 is the velocity of the wave.

Exercise 1.2: Consider an initial condition $v(x,0) = \exp(-x^2)$ and find an implicit solution of (1.3c) in terms of the Fourier spectrum of the initial condition.

Ans: $v(x,t) = \frac{1}{2\sqrt{\pi}}\int_{-\infty}^{\infty} e^{-\left(\frac{k}{2}\right)^2} e^{ik(x+k^2 t)}\,dk$.

Exercise 1.3: The integral of the implicit solution, derived in Exercise 1.2, is evaluated numerically; show two conservation conditions of the solution, which are imposed in numerical integration.

Ans: $\int_{-\infty}^{\infty} v(x,t)dx = \sqrt{\pi}$ and $\int_{-\infty}^{\infty} v^2(x,t)dx = \sqrt{\pi/2}$.

With the aid of $v_{xt} = v_{tx}$ and $u_{xt} = u_{tx}$, a time (t) derivative of (1.3a) and a spatial (x) derivative of (1.3b) are combined as

$$uu_t = 3u_x u_{xx} - uu_{xxx} \tag{1.3d}$$

Therefore, it requires that $u_x u_{xx} = -uu_x^2$, i.e., $\left(u_x + \frac{1}{2}u^2\right)_x = 0 = v_{xx}$. It is then obtained that $v_t = 0$ and $v(t,x) = \beta x + C$, where β is called the Bäcklund parameter and C is a constant; and (1.3a) and (1.3b) reduce to

$$u_x = \beta - \frac{1}{2}u^2 \tag{1.3e}$$

$$u_t = uu_{xx} - 2u_x^2 \tag{1.3f}$$

With the aid of $u_{xt} = u_{tx}$ (1.3e) and (1.3f) yield

$$-uu_t = uu_{xxx} - 3u_x u_{xx} = uu_{xxx} + 3uu_x^2$$

which turns to (1.2a). The relationship, $u_{xx} = -uu_x$ implied by (1.3e), is employed.

Let $u = \sqrt{2\beta}\psi$, (1.3e) becomes $\psi_x = \sqrt{\frac{\beta}{2}}(1 - \psi^2)$; in the two cases of $\psi^2 \leq 1$ and ≥ 1, it is integrated to give two solutions $u_{1,2}$ of (1.2a) as

$$u_1(x,t) = \sqrt{2\beta}\psi_1 = \sqrt{2\beta}\tanh\left[\sqrt{\frac{\beta}{2}}(x - \alpha t) + \theta_0\right]$$

$$\text{for } \psi_1^2 \leq 1 \tag{1.3g}$$

and

$$u_2(x,t) = \sqrt{2\beta}\psi_2 = \sqrt{2\beta}\coth\left[\sqrt{\frac{\beta}{2}}(x - \alpha t) + \theta_0\right]$$

$$\text{for } \psi_2^2 \geq 1 \tag{1.3h}$$

where the velocity, $\alpha = 2\beta$, is determined by substituting (1.3g) and (1.3h) into (1.3f); (1.3h) is a singular solution. Then, a soliton

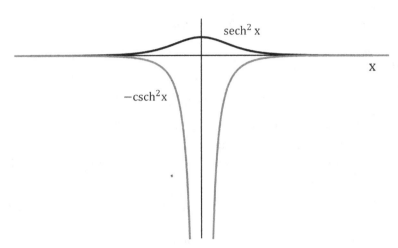

Fig. 1.3. Representation of a soliton and a singular solution of the KdV equation (1.1).

solution ϕ_1 and a singular solution ϕ_2 of (1.1) are obtained as

$$\phi_1(x,t) = u_{1_x} = \beta \operatorname{sech}^2 \left[\sqrt{\frac{\beta}{2}}(x - 2\beta t) + \theta_0 \right] \tag{1.4a}$$

and

$$\phi_2(x,t) = u_{2_x} = -\beta \operatorname{csch}^2 \left[\sqrt{\frac{\beta}{2}}(x - 2\beta t) + \theta_0 \right] \tag{1.4b}$$

Plots of the two solutions are presented in Fig. 1.3.

Exercise 1.4: Set $u = -\sqrt{2\beta}\psi$ in (1.3e) and find solutions of (1.1).

Ans: Two solutions are the same as (1.4a) and (1.4b).

It is noted that both $\pm v$ are the solutions of (1.3c). Specifically, (1.3a) (as well as (1.3b)) can relate a single solution v of (1.3c) to two distinct separate solutions u_1 and u_2 of (1.2a), e.g., $v_x = u_{1_x} + \frac{1}{2}u_1^2$ and $-v_x = u_{2_x} + \frac{1}{2}u_2^2$. However, (1.3a) transforms u to v; it is not straightforward to combine u_1 and u_2 to a form only involving v. On the other hand, a transform from v to u can render useful combinations of u_1 and u_2 for investigating the nonlinear superposition method, which generates multiple-solitons solutions.

B. Derivation of Bäcklund transform of reverse form

First, take a subtraction of the u_1 and u_2 equations set up by (1.2a) and set $u_1 - u_2 = 2v$, it yields

$$v_t + v_{xxx} + 3v_x(u_1 + u_2)_x = 0 \tag{1.5a}$$

To achieve that both $\pm v$ are the solutions of (1.5a), $(u_1 + u_2)_x$ needs to be an even function of v. Thus, $(u_1 + u_2)_x = \beta + vf(v)$ is assumed, where β is a constant parameter and $f(v)$ is an odd function of v. Then, (1.5a) becomes

$$v_t + v_{xxx} + 3v_x(\beta + vf) = 0 \tag{1.5b}$$

Next, add the u_1 and u_2 equations set up by (1.2a) and take a derivative of x on the resulting equation, it yields

$$v_t + \frac{v}{f}f_t + v_{xxx} + \frac{v}{f}f_{xxx} + \frac{3}{f}(v_{xx}f_x + f_{xx}v_x)$$

$$+ \frac{3}{f}[4v_x v_{xx} + (\beta + vf)(vf_x + fv_x)] = 0 \tag{1.5c}$$

Since (1.5c) should reduce to (1.5b), $f = -2v$ is obtained. Then,

$$u_{1_x} = v_x + \frac{\beta}{2} - v^2 \quad \text{and} \quad u_{2_x} = -v_x + \frac{\beta}{2} - v^2 \tag{1.5d}$$

The equation to transform v to u is derived as

$$u_x = v_x + \frac{\beta}{2} - v^2 \tag{1.6a}$$

$$u_t = v_t + (2vv_{xx} - v_x^2) - \frac{3}{4}(\beta - 2v^2)^2 \tag{1.6b}$$

As $\beta = 0$, (1.5b) becomes the modified KdV (mKdV) equation, and (1.6a) is called the Miura transformation, which transforms KdV equation to mKdV equation as discussed in Section 3.2.2 of Chapter 3.

Exercise 1.5: Verify that the pair of Bäcklund transform (1.6a) and (1.6b) represents the KdV equation (1.1).

Ans: Apply the relations $v_{xt} = v_{tx}$ and $u_{xt} = u_{tx}$.

Consequently, an equation relating $(u_1 + u_2)$ and $(u_1 - u_2)$ is also obtained as

$$(u_1 + u_2)_x = \beta - 2v^2 = \beta - \frac{1}{2}(u_1 - u_2)^2 \tag{1.7a}$$

Furthermore, (1.5b) can be re-expressed as

$$v_t + v_{xxx} + 3(\beta - 2v^2)v_x$$

$$= (u_1 - u_2)_t + (u_1 - u_2)_{xxx} + 3\beta(u_1 - u_2)_x - \frac{1}{2}[(u_1 - u_2)^3]_x$$

$$= 0 \tag{1.7b}$$

Though (1.7b) is different from (1.3c), both equations have the feature of $\pm v$ as solutions. In case of $u_2 = 0$, (1.7a) and (1.7b) reduce to

$$u_{1_x} = \beta - \frac{1}{2}u_1^2 \tag{1.7c}$$

$$u_{1_t} = -u_{1_{xxx}} - 3\beta u_{1_x} + \frac{3}{2}u_1^2 u_{1_x}$$

$$= (u_{1_x})^2 + u_1 u_{1_{xx}} - 3\left(u_{1_x} + \frac{1}{2}u_1^2\right)u_{1_x} + \frac{3}{2}u_1^2 u_{1_x}$$

$$= u_1 u_{1_{xx}} - 2(u_{1_x})^2 \tag{1.7d}$$

These are the same as the set of Bäcklund transform (1.3e) and (1.3f). Thus, $u_1(= u)$ and $\phi_1(= \phi = u_x)$ are given by (1.3g) and (1.4a), respectively.

The pair of Bäcklund transform (1.6a) and (1.6b) is applied to find solitary solutions of the KdV equation (1.1). Set $v_{1,2} = \pm v$ as two solutions of (1.7b), which are transformed into $u_{1,2}$, two solutions of (1.2a), then the corresponding solutions of the KdV equation (1.1) are given as $\phi_{1,2} = (u_{1,2})_x$. Since a constant satisfies (1.2a), $u_{2_x} = 0$ is chosen, then (1.6a) becomes

$$v_x = \frac{\beta}{2} - v^2$$

This equation is the same as (1.3e); again, in the two cases of $|v| \lessgtr \sqrt{\frac{\beta}{2}}$, it is integrated to obtain $v^{(1)} = \sqrt{\frac{\beta}{2}} \tanh\left[\sqrt{\frac{\beta}{2}}x + \theta(t)\right] = -v_2^{(1)}$

and $v^{(2)} = \sqrt{\frac{\beta}{2}} \coth\left[\sqrt{\frac{\beta}{2}}x + \theta(t)\right] = -v_2^{(2)}$. Now, substitute $v_2^{(1,2)}$ into (1.6b) with $u_{2_t} = 0$, an equation for $\theta(t)$, in both cases, is derived as

$$\theta_t = -4\left(\frac{\beta}{2}\right)^{\frac{3}{2}}$$

It is integrated into $\theta(t) = -4\left(\frac{\beta}{2}\right)^{\frac{3}{2}}t + \theta_0$. Thus, a solitary solution $\phi_1 = (u_1^{(1)})_x$ and a singular solution $\phi_2 = (u_1^{(2)})_x$ are obtained, through the transform (1.6a), as

$$\phi_1 = (v^{(1)})_x + \frac{\beta}{2} - (v^{(1)})^2 = \beta \operatorname{sech}^2\left[\sqrt{\frac{\beta}{2}}(x-2\beta t) + \theta_0\right]$$

and

$$\phi_2 = (v^{(2)})_x + \frac{\beta}{2} - (v^{(2)})^2 = -\beta \operatorname{csch}^2\left[\sqrt{\frac{\beta}{2}}(x - 2\beta t) + \theta_0\right]$$

which is the same as (1.4a) and (1.4b), respectively.

Exercise 1.6: Set $u_{1_x} = 0$ in (1.6a) and find solutions of (1.1) with the aid of the Bäcklund transform (1.6a) and (1.6b).

Ans: Two solutions are the same as (1.4a) and (1.4b).

Exercise 1.7: Equation (1.4a) indicates that $\phi_1 = 2\operatorname{sech}^2(x - 4t)$ is a solution of the KdV equation (1.1). Set $u_{1_x} = \phi_1 = 2\operatorname{sech}^2(x - 4t)$ and $\beta = 0$ in (1.6a), to find the corresponding solution v_1 of the mKdV equation and then to generate a second solution ϕ_2 of the KdV equation.

Ans: $v_1 = -\frac{\operatorname{sech}^2(x-4t)}{\tan(x-4t)}$; $\phi_2 = -2\operatorname{csch}^2(x - 4t)$.

1.3.2 *Burgers equation*

$$\frac{\partial}{\partial t}\phi + \phi\frac{\partial}{\partial x}\phi = b\frac{\partial^2}{\partial x^2}\phi \tag{1.8a}$$

Burgers equation mimics the Navier–Stokes equations of fluid motion through its fluid-like expressions for nonlinear advection and linear diffusion. While nonlinearity would steepen fronts into discontinuous shocks, diffusion would smooth them away. It has been applied to describe the behavior of shock waves, traffic flow, and acoustic transmission.

Multiply (1.8a) with a function v, it yields

$$
(v\phi)_t - \phi v_t + \left[v \left(\frac{\phi^2}{2} - b\phi_x \right) \right]_x - \left(\frac{\phi^2}{2} - b\phi_x \right) v_x = 0
$$

$$
= (v\phi)_t - b \left(v\phi_x - \frac{v\phi^2}{2b} \right)_x - \phi \left(v_t + \frac{\phi v_x}{2} - b\frac{\phi_x v_x}{\phi} \right) \quad (1.8b)
$$

A Bäcklund transform to relate v and ϕ is introduced as

$$
v_x = -\frac{v\phi}{2b} \quad (1.8c)
$$

$$
v_t = -\frac{\phi v_x}{2} + b\frac{\phi_x v_x}{\phi} \quad (1.8d)
$$

Then, (1.8b), with the aid of $v\phi = -2bv_x$, becomes

$$
v_t - bv_{xx} = 0 \quad (1.8e)
$$

It shows that the viscid Burgers equation (1.8a) is transformed into a linear diffusion equation (heat equation) (1.8e), which can be solved analytically as shown in Section 3.3 of Chapter 3. Given a solution of the linear diffusion equation, a solution to Burgers equation is obtained via the transform (1.8c), which is re-expressed as $\phi = -2bv_x/v$. This relationship is called "Cole–Hopf transformation".

Exercise 1.8: Find the solution of the linear diffusion equation (1.8e) with the initial condition $v(x,0) = \exp(-x^2)$.

Ans: $v(x,t) = \frac{1}{\sqrt{1+4bt}} \exp\left(-\frac{x^2}{1+4bt}\right)$

Exercise 1.9: Show that $v(x,t) = C + \frac{A}{\sqrt{t}} \exp\left(-\frac{x^2}{4bt}\right)$ is a solution of (1.8e), where A and C are constants.

1.3.3   *The sine-Gordon equation*

$$\varphi_{tt} = a\varphi_{xx} + b\sin\lambda\varphi \tag{1.9}$$

The sine-Gordon equation appears in several physical applications, including the propagation of fluxions in Josephson junctions (a junction between two superconductors), the motion of rigid pendula attached to a stretched wire, and dislocations in crystals. It is a nonlinear hyperbolic partial differential equation, where $\varphi = \varphi(x,t)$. In the small-amplitude case ($\sin\lambda\varphi \approx \lambda\varphi$), it reduces to the Klein–Gordon equation

$$\varphi_{tt} - a\varphi_{xx} - b\lambda\varphi = 0$$

Transform (1.9) to the light-cone coordinates (τ, z), where

$$\tau = \sqrt{-\frac{b}{a}}\,\frac{x + \sqrt{a}t}{2} \quad \text{and} \quad z = \sqrt{-\frac{b}{a}}\,\frac{x - \sqrt{a}t}{2}$$

then, (1.9) takes the form

$$\varphi_{\tau z} = \sin\lambda\varphi \tag{1.10}$$

It is now to introduce a function ϑ and to rearrange (1.10) as

$$(\vartheta + \varphi)_{\tau z} - (\vartheta - \varphi)_{\tau z}$$

$$= 2\sin\lambda\frac{(\vartheta+\varphi)}{2}\cos\lambda\frac{(\vartheta-\varphi)}{2} - 2\cos\lambda\frac{(\vartheta+\varphi)}{2}\sin\lambda\frac{(\vartheta-\varphi)}{2}$$

$$= \frac{4\sin\lambda\frac{(\vartheta+\varphi)}{2}}{\lambda\,(\vartheta-\varphi)_\tau}\frac{\partial}{\partial\tau}\sin\lambda\frac{(\vartheta-\varphi)}{2}$$

$$- \frac{4\sin\lambda\frac{(\vartheta-\varphi)}{2}}{\lambda\,(\vartheta+\varphi)_z}\frac{\partial}{\partial z}\sin\lambda\frac{(\vartheta+\varphi)}{2}$$

An auto-Bäcklund transform of φ and ϑ is then given by

$$\vartheta_\tau = \varphi_\tau + 2\frac{\alpha}{\sqrt{\lambda}}\sin\lambda\left(\frac{\vartheta+\varphi}{2}\right) \tag{1.11a}$$

$$\vartheta_z = -\varphi_z + \frac{2}{\alpha\sqrt{\lambda}}\sin\lambda\left(\frac{\vartheta-\varphi}{2}\right) \tag{1.11b}$$

where ϑ is also a solution of (1.10), i.e., $\vartheta_{\tau z} = \sin \lambda \vartheta$, and α is called the Bäcklund parameter.

Let $f = \lambda \frac{\vartheta + \varphi}{2}$ and $g = \lambda \frac{\vartheta - \varphi}{2}$ then (1.11a) and (1.11b) become

$$g_\tau = \alpha \sqrt{\lambda} \sin f \quad \text{and} \quad f_z = \frac{\sqrt{\lambda}}{\alpha} \sin g$$

One now looks for a traveling wave solution, i.e.,

$$f(\tau, z) = f(\eta) \quad \text{and} \quad g(\tau, z) = g(\eta)$$

where $\eta = kz + \mu\tau + \theta_0$.

Thus,

$$g_\eta = \frac{\alpha \sqrt{\lambda}}{\mu} \sin f \quad \text{and} \quad f_\eta = \frac{\sqrt{\lambda}}{\alpha k} \sin g$$

The system of equations has a solution $f = -g$, with $\alpha^2 = \frac{\mu}{k}$, i.e., $\vartheta = 0$ and $\varphi = 2f/\lambda$; it leads to

$$f_\eta = \pm \frac{\sqrt{\lambda}}{\sqrt{\mu k}} \sin f \tag{1.12}$$

This equation is integrated as $\ln \tan \frac{f}{2} = \pm \frac{\sqrt{\lambda}}{\sqrt{\mu k}} \eta$, which becomes

$$f(\eta) = 2 \tan^{-1} e^{\pm \frac{\sqrt{\lambda}}{\sqrt{\mu k}} \eta} \tag{1.13}$$

Two analytical solutions of the sine-Gordon equation are then obtained as

$$\varphi_\pm(\eta) = \frac{4}{\lambda} \tan^{-1} e^{\pm \frac{\sqrt{\lambda}}{\sqrt{\mu k}} \eta} \tag{1.14a}$$

These represent "kink $(+)$" and "anti-kink $(-)$" solitons, as shown in Fig. 1.4.

Convert η to the original variables (x, t),

$$\eta = \sqrt{-\frac{b}{a} \frac{k}{2}} (x - \sqrt{a} t) + \sqrt{-\frac{b}{a} \frac{\mu}{2}} (x + \sqrt{a} t) + \theta_0$$

$$= \sqrt{-\frac{b}{a}} \left(\frac{k + \mu}{2} \right) x - \sqrt{-b} \left(\frac{k - \mu}{2} \right) t + \theta_0$$

$$= k' x + \mu' t + \theta_0$$

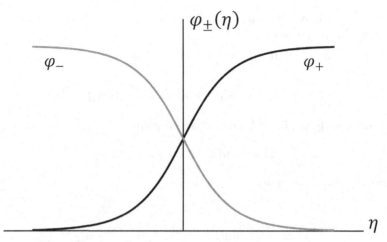

Fig. 1.4.   Representation of the kink (+) and anti-kink (−) solutions of the sine-Gordon equation (1.10).

where $k' = \frac{1}{2}\sqrt{-\frac{b}{a}}(k + \mu)$ and $\mu' = -\frac{1}{2}\sqrt{-b}(k - \mu)$ and then, $\mu k = \frac{1}{b}(\mu'^2 - ak'^2)$.

The solutions $\varphi_\pm(x, t)$ of (1.9) are obtained as

$$\varphi_\pm(x, t) = \frac{4}{\lambda}\tan^{-1}e^{\pm\frac{\sqrt{\lambda}}{\sqrt{\mu k}}\eta}$$

$$= \frac{4}{\lambda}\tan^{-1}\exp \pm \frac{b\lambda(k'x + \mu't + \theta_0)}{\sqrt{b\lambda\left(\mu'^2 - ak'^2\right)}}$$

$$\text{for } b\lambda(\mu'^2 - ak'^2) > 0 \qquad (1.14b)$$

Exercise 1.10: Verify that (1.14b) satisfies (1.9).

1.3.4   *The Liouville equation*

$$\left(\frac{\partial^2}{\partial x^2} + \frac{\partial^2}{\partial y^2}\right)\ln f(x, y) = -Kf^2 = -K\exp(2\ln f) \qquad (1.15)$$

The Liouville equation describes the evolution of the phase space distribution function for the conservative Hamiltonian system. In essence, it is a continuity equation for the flux (evolution of ensemble in the phase space, where the number of systems in the ensemble is constant).

Set $u = 2 \ln f$, (1.15) becomes

$$\left(\frac{\partial^2}{\partial x^2} + \frac{\partial^2}{\partial y^2} \right) u(x, y) = -2K \exp u \tag{1.16}$$

With the coordinate transform

$$\eta = i\sqrt{\frac{K}{2}}(x + iy) \quad \text{and} \quad \xi = i\sqrt{\frac{K}{2}}(x - iy) \tag{1.17}$$

Equation (1.16) reduces to

$$u_{\eta\xi} = \exp u \tag{1.18}$$

With the aid of a function v, (1.18) is re-expressed as

$$(v + u)_{\eta\xi} - (v - u)_{\eta\xi} = 2 \exp \left(\frac{v + u}{2} \right) \exp \left[- \left(\frac{v - u}{2} \right) \right]$$

$$= 2 \left[\frac{\partial}{\partial \xi} \exp \left(\frac{v + u}{2} \right) \right] \frac{\exp \left[- \left(\frac{v-u}{2} \right) \right]}{(v + u)_\xi}$$

$$- 2 \frac{\exp \left(\frac{v+u}{2} \right)}{(v - u)_\eta} \left\{ \frac{\partial}{\partial \eta} \exp \left[- \left(\frac{v - u}{2} \right) \right] \right\}$$

A Bäcklund transform of u and v is found as

$$v_\eta = u_\eta + 2a \exp \left(\frac{v + u}{2} \right) \tag{1.19a}$$

$$v_\xi = -u_\xi - \frac{1}{a} \exp \left[- \left(\frac{v - u}{2} \right) \right] \tag{1.19b}$$

where a is an arbitrary parameter, and if u is a solution of (1.18), i.e., $u_{\eta\xi} = \exp u$, then v is a solution of the much simpler equation, $v_{\eta\xi} = 0$, and vice versa.

Thus, $v = A\eta + B\xi + C$ and (1.19a) and (1.19b) become

$$A = u_\eta + 2a \exp \left(\frac{u + A\eta}{2} \right) \exp \left(\frac{B\xi + C}{2} \right) \tag{1.20a}$$

$$B = -u_\xi - \frac{1}{a} \exp \left(\frac{u - B\xi}{2} \right) \exp \left(-\frac{A\eta + C}{2} \right) \tag{1.20b}$$

It shows that one can then solve the (second-order partial differential) Liouville equation (1.18) by working with two much simpler (first-order partial differential) equations.

Set $F = \frac{u+A\eta}{2}$ and $G = \frac{u-B\xi}{2}$ then (1.20a) and (1.20b) become

$$F_\eta = A[1 - P(\xi)e^F] \tag{1.21a}$$

$$G_\xi = -B[1 + Q(\eta)e^G] \tag{1.21b}$$

where $P(\xi) = \frac{a}{A}\exp\left(\frac{B\xi+C}{2}\right)$ and $Q(\eta) = \frac{1}{2aB}\exp\left(-\frac{A\eta+C}{2}\right)$.
Equations (1.21a) and (1.21b) are integrated as

$$\ln[e^{-F} - P(\xi)] = -A\eta + \ln T(\xi) \quad \text{and} \quad \ln[e^{-G} + Q(\eta)]$$
$$= B\xi + \ln W(\eta)$$

which are converted to

$$e^{-F} - P(\xi) = T(\xi)e^{-A\eta} \quad \text{and} \quad e^{-G} + Q(\eta) = W(\eta)e^{B\xi}$$

Substitute $F = \frac{u+A\eta}{2}$ and $G = \frac{u-B\xi}{2}$ back into these two results, it yields

$$e^{-\frac{u}{2}} = P(\xi)e^{\frac{A\eta}{2}} + T(\xi)e^{-\frac{A\eta}{2}} = -Q(\eta)e^{-\frac{B\xi}{2}} + W(\eta)e^{\frac{B\xi}{2}}$$

It is re-expressed and sets up an equation

$$\left[\frac{a}{A}\exp\left(\frac{A\eta+C}{2}\right) - W(\eta)\right]e^{\frac{B\xi}{2}}$$
$$+ \left[T(\xi) + \frac{1}{2aB}\exp\left(-\frac{B\xi+C}{2}\right)\right]e^{-\frac{A\eta}{2}} = 0$$

It implies that

$$T(\xi) = -\frac{1}{2aB}\exp\left[-\frac{1}{2}(B\xi + C)\right] \quad \text{and}$$

$$W(\eta) = \frac{a}{A}\exp\left[\frac{1}{2}(A\eta + C)\right]$$

which lead to

$$u = -2\ln\left(\frac{a}{A}e^{\frac{v}{2}} - \frac{1}{2aB}e^{-\frac{v}{2}}\right) \tag{1.22}$$

Convert (η, ξ) to (x, y)

$$v(\eta, \xi) = A\eta + B\xi + C = i\sqrt{\frac{K}{2}}[A(x + iy) + B(x - iy)] + C = v(x, y)$$

A solution of (1.15) is obtained as

$$f(x, y) = \exp\frac{u}{2} = \frac{1}{\frac{a}{A}e^{\frac{v}{2}} - \frac{1}{2aB}e^{-\frac{v}{2}}} = \frac{1}{\frac{\alpha}{A}e^{H} - \frac{1}{2\alpha B}e^{-H}} \tag{1.23a}$$

where $e^{\frac{v}{2}} = e^{\frac{C}{2}}\exp\left\{i\sqrt{\frac{K}{8}}[A(x + iy) + B(x - iy)]\right\} = e^{\frac{C}{2}}\exp\left\{i\sqrt{\frac{K}{8}}[(A+B)x + i(A-B)y]\right\}$, $\alpha = ae^{C/2}$, and $H(x, y) = i\sqrt{\frac{K}{8}}[(A+B)x + i(A-B)y]$.

In the limit A and $B \to 0$, i.e., $v = C$; set $B = \frac{A}{2\alpha^2}$, (1.23a) reduces to

$$f = -i\sqrt{\frac{2}{K}}\frac{1}{\left(\alpha + \frac{1}{2\alpha}\right)x + i\left(\alpha - \frac{1}{2\alpha}\right)y} \tag{1.23b}$$

Exercise 1.11: Show that (1.16) converts to (1.18) with the coordinate transform (1.17).

Exercise 1.12: Derive (1.22) with the aid of the relations $T(\xi)$ and $W(\eta)$.

1.3.5 *Cubic nonlinear Schrödinger equation*

$$-\frac{1}{2}\frac{\partial^2}{\partial\xi^2}\varphi - \alpha|\varphi|^2\varphi = i\frac{\partial}{\partial\tau}\varphi \tag{1.24}$$

The nonlinear Schrödinger equation (NLSE) models the slowly varying envelope dynamics of a weakly nonlinear quasi-monochromatic wave packet in dispersive media. It arises in various physical contexts in the description of nonlinear waves, such as the propagation of a laser beam in a medium whose index of reflection is sensitive to the wave amplitude, water waves at the free surface of an ideal, and plasma waves. Some applications include optical communications, laser surgeries, and remote sensing.

Set $\varphi(\xi,\tau) = \phi(x,t), t = \frac{\tau}{2}, x = \xi$, and $\alpha = 1$, (1.24) is re-expressed as

$$i\phi_t + \phi_{xx} + 2|\phi|^2\phi = 0 \tag{1.25a}$$

where partial derivatives are denoted by subscripts. Since $\phi(x,t)$ is a complex function, ϕ and ϕ^* are transformed separately into Θ and Ψ via Bäcklund transforms

$$\Theta_x = (\phi_x + i\lambda\phi)e^{i\lambda(x+2\lambda t)} = \frac{\partial}{\partial x}[\phi e^{i\lambda(x+2\lambda t)}]$$

$$\Theta_t = \left(\phi_t + 2i\lambda^2\phi\right)e^{i\lambda(x+2\lambda t)} = \frac{\partial}{\partial t}[\phi e^{i\lambda(x+2\lambda t)}]$$

and

$$\Psi_x = \left(\phi_x^* - i\lambda\phi^*\right)e^{-i\lambda(x+2\lambda t)} = \frac{\partial}{\partial x}[\phi^* e^{-i\lambda(x+2\lambda t)}]$$

$$\Psi_t = \left(\phi_t^* - 2i\lambda^2\phi^*\right)e^{-i\lambda(x+2\lambda t)} = \frac{\partial}{\partial t}[\phi^* e^{-i\lambda(x+2\lambda t)}]$$

In essence,

$$\Theta = \phi e^{i\lambda(x+2\lambda t)} \quad \text{and} \quad \Psi = \phi^* e^{-i\lambda(x+2\lambda t)} \tag{1.25b}$$

Substitute $\phi = \Theta e^{-i\lambda(x+2\lambda t)}$ into (1.25a) and $\phi^* = \Psi e^{i\lambda(x+2\lambda t)}$ into the complex conjugate of (1.25a), it yields

$$i(\Theta_t - 2i\lambda^2\Theta) + \Theta_{xx} - 2i\lambda\Theta_x - \lambda^2\Theta + 2|\phi|^2\Theta$$
$$= (\Theta_x + i\lambda\Theta - \phi\Psi)_x - i\lambda(\Theta_x + i\lambda\Theta - \phi\Psi)$$
$$+ \phi(\Psi_x - i\lambda\Psi + \phi^*\Theta) + i(\Theta_t - 2\lambda\Theta_x - i|\phi|^2\Theta - i\phi_x\Psi) = 0$$

and

$$-i(\Psi_t + 2i\lambda^2\Psi) + \Psi_{xx} + 2i\lambda\Psi_x - \lambda^2\Psi + 2|\phi|^2\Psi$$
$$= (\Psi_x - i\lambda\Psi + \phi^*\Theta)_x + i\lambda(\Psi_x - i\lambda\Psi + \phi^*\Theta)$$
$$- \phi^*(\Theta_x + i\lambda\Theta - \phi\Psi) - i(\Psi_t - 2\lambda\Psi_x + i|\phi|^2\Psi - i\phi_x^*\Theta) = 0$$

Thus, (Θ, Ψ) are governed by the following sets of coupled first-order partial differential equations

$$\begin{bmatrix} \Theta \\ \Psi \end{bmatrix}_x = \begin{bmatrix} -i\lambda & \phi \\ -\phi^* & i\lambda \end{bmatrix} \begin{bmatrix} \Theta \\ \Psi \end{bmatrix} \tag{1.26}$$

$$\begin{bmatrix} \Theta \\ \Psi \end{bmatrix}_t = \begin{bmatrix} -2i\lambda^2 + i|\phi|^2 & i\phi_x + 2\lambda\phi \\ i\phi_x^* - 2\lambda\phi^* & 2i\lambda^2 - i|\phi|^2 \end{bmatrix} \begin{bmatrix} \Theta \\ \Psi \end{bmatrix} \tag{1.27}$$

With the aid of $\Theta_{xt} = \Theta_{tx}$ (or $\Psi_{xt} = \Psi_{tx}$), it can be shown that (1.26) and (1.27) imply (1.25a).

Exercise 1.13: Verify that the pair of transforms (1.26) and (1.27) represent the cubic nonlinear Schrödinger equation (1.25a).

Next, substitute these relations $\Theta = \phi e^{i\lambda(x+2\lambda t)}$ and $\Psi = \phi^* e^{-i\lambda(x+2\lambda t)}$ back into (1.26) and (1.27), a pair of first-order differential equations for the $\phi(x,t)$ is obtained as

$$\phi_x + 2i\lambda\phi = |\phi|^2 e^{-i2\lambda(x+2\lambda t)} \tag{1.28}$$

and

$$\phi_t + 4i\lambda^2\phi = i|\phi|^2\phi + (i\phi_x + 2\lambda\phi)\phi^* e^{-i2\lambda(x+2\lambda t)}$$
$$= i|\phi|^2 [\phi + \phi^* e^{-i4\lambda(x+2\lambda t)}] + 4\lambda|\phi|^2 e^{-i2\lambda(x+2\lambda t)} \tag{1.29}$$

Introduce $V = \phi e^{i2\lambda(x+2\lambda t)}$, (1.28) and (1.29) become

$$V_x = |\phi|^2 = |V|^2 |e^{-i2\lambda(x+2\lambda t)}|^2 \tag{1.30a}$$

and

$$V_t = 4\lambda|\phi|^2 + i|u|^2 [\phi e^{i2\lambda(x+2\lambda t)} + \phi^* e^{-i2\lambda(x+2\lambda t)}]$$
$$= \{4\lambda + i[\phi e^{i2\lambda(x+2\lambda t)} + \phi^* e^{-i2\lambda(x+2\lambda t)}]\} V_x \tag{1.31a}$$

Equation (1.30a) indicates that V is a real function. Set $\lambda = \beta + i\kappa$, (1.30a) and (1.31a) become

$$V_x = |\phi|^2 = |V|^2 e^{4\kappa(x+4\beta t)} \tag{1.30b}$$

and

$$V_t = [4(\beta + i\kappa) + i(V + V^* e^{4\kappa(x+4\beta t)})] V_x \tag{1.31b}$$

The imaginary part and real part of (1.31b) give

$$4\kappa + [V + V^*e^{4\kappa(x+4\beta t)}] = 4\kappa + V(1 + e^{4\kappa(x+4\beta t)}) = 0$$

and

$$V_t = 4\beta V_x$$

Thus,

$$V(x,t) = V(x + 4\beta t)$$

and

$$V = -\frac{4\kappa}{1 + e^{4\kappa(x+4\beta t)}} = \phi e^{i2[\beta x + 2(\beta^2 - \kappa^2)t]}e^{-2\kappa(x+4\beta t)}$$

Then,

$$\phi(x,t) = -2\kappa \operatorname{sech} 2\kappa(x + 4\beta t)e^{-i2[\beta x + 2(\beta^2 - \kappa^2)t]} \tag{1.32a}$$

where β and κ can be positive or negative real values. Equation (1.32a) represents a traveling solitary wave. In the case of $\beta = 0$, (1.32a) reduces to a stationary soliton

$$\phi_s(x,t) = -2\kappa \operatorname{sech} 2\kappa x \; e^{i4\kappa^2 t} \tag{1.32b}$$

Exercise 1.14: Verify that (1.32b) satisfies the cubic nonlinear Schrödinger equation (1.25a).

Ans: This is achieved with the aid of $\frac{d}{dx} \operatorname{sech} 2\kappa x = -2\kappa \operatorname{sech} 2\kappa x \tanh 2\kappa x$ and $\frac{d^2}{dx^2} \operatorname{sech} 2\kappa x = 4\kappa^2 \operatorname{sech} 2\kappa x(1 - 2 \operatorname{sech}^2 2\kappa x)$.

Problems

P1.1. Traveling wave approach converts partial differential equation (PDE) to ordinary differential equation (ODE); certain nonlinear equations may be solved via this approach to obtain analytical stationary solutions. Consider generalized Fisher's equation

$$u_t - Du_{xx} - ru(1 - u^q) = 0 \tag{P1.1}$$

Set $u(x,t) = u(\eta)$, where $\eta = kx - \omega t$ and $\frac{\omega}{k} = V$ is the wave velocity, and introduce a transformation $u = Q^2$, where $Q = \left(\frac{1}{1+e^\eta}\right)^{1/q} = \frac{1}{2^{1/q}}\left(1 - \tanh \frac{\eta}{2}\right)^{1/q}$.

(1) Show that an analytical solution of (P1.1) is given as

$$u = \frac{1}{2^{\frac{2}{q}}} \left(1 - \tanh \frac{\eta}{2}\right)^{\frac{2}{q}}$$

$$= \frac{1}{2^{\frac{2}{q}}} \left[1 \mp \tanh \sqrt{\frac{r}{2D(q+2)}} \frac{q}{2}\right.$$

$$\left. \times \left(x \mp \sqrt{\frac{rD}{2(q+2)}}(q+4)t\right)\right]^{\frac{2}{q}}$$

(2) Find the solution of the Fisher equation (i.e., $q = 1$).

(3) Find the solution of the Burgers–Huxley equation ($D = r = 1$ and $q = 2$).

P1.2. Find an analytical traveling wave solution of the Boussinesq equation

$$u_{tt} - u_{xx} - \beta(u^2)_{xx} - \alpha u_{xxxx} = 0 \qquad (\text{P1.2})$$

with the aid of the transformation $u(\eta) = P + Q \tanh^2 \eta$, where $\eta = k(x - Vt)$ and V is the wave velocity.

P1.3. The Dym equation

$$u_t = u^3 u_{xxx} \qquad (\text{P1.3})$$

represents a system in which dispersion and nonlinearity are coupled together. Show that a traveling wave solution is given as

$$u(x, t) = [3\alpha(x + 4\alpha^2 t)]^{\frac{2}{3}}$$

P1.4. The Dym equation (P1.3) has strong links to the Korteweg–de Vries (KdV) and modified Korteweg–de Vries (mKdV) equations. Show that with the coordinate transform

$$y = -\int \frac{1}{u(x,t)} dx \quad \text{and} \quad \tau = t$$

and the Cole–Hopf transformation

$$\psi(y, \tau) = \frac{1}{2}\frac{u_y}{u} = -\frac{1}{2}u_x$$

The Dym equation $u_t = u^3 u_{xxx}$ is transformed into the mKdV equation

$$\psi_\tau + \psi_{yyy} - 6\psi^2 \psi_y = 0$$

P1.5. The Dym equation (P1.3) has a solution

$$u(x,t) = [3\alpha(x + 4\alpha^2 t)]^{\frac{2}{3}}$$

Apply the transformations to find the corresponding solution of the mKdV equation.

P1.6. Find the corresponding solution of the KdV equation to the mKdV equation solution of P1.5.

P1.7. Consider a family of the KdV equations represented by a generic nonlinear partial differential equation

$$\frac{\partial}{\partial t}\phi + \frac{\partial^3}{\partial x^3}\phi + \alpha_p \phi^p \frac{\partial}{\partial x}\phi = 0 \qquad (P1.4)$$

where the integer $p = 1, 2, 3, \ldots$ leads (P1.4) to the KdV, mKdV, super mKdV, ... equations.

Find soliton solutions of (P1.4) via traveling wave approach; set $\phi(x,t) = \phi(\xi)$ where $\xi = x - Vt$ for a constant speed V and assume the ansatz $\phi = A_p \operatorname{sech}^{2/p}\xi$.

P1.8. Solve the nonlinear Schrödinger equation (1.25a) via the traveling wave approach by introducing the ansatz $\phi(x,t) = A \operatorname{sech}\xi \, e^{iB\eta}$, where $\xi = x - V_s t$ and $\eta = x - St$.

Chapter 2

Formulation of Nonlinear Wave Equations in Plasma

Although the propagation of different nonlinear waves was studied originally in different media, many features of wave propagation in plasma can mimic most of the nonlinear wave phenomena. This is because plasma is a nonlinear dispersive dielectric medium; it supports high-frequency EM and electron plasma (Langmuir) waves as well as low-frequency ion acoustic waves. High-frequency wave packets contain carriers; the nonlinear wave phenomena are revealed in the evolution of the modulation amplitudes (envelopes), which are governed by the nonlinear Schrödinger equation (NLSE). The baseband low-frequency waves are governed by the Korteweg–de Vries (KdV) equation in the study of nonlinear dispersive wave phenomena in collisionless plasma and by the Burgers equation in the study of nonlinear dissipative wave phenomena in collision plasma. These equations are derived in the following. In Section 1.3 of Chapter 1, Bäcklund transform technique is introduced to find stationary solutions of these equations. A selective overview of the characteristic features of the solutions of these nonlinear equations and of the nonlinear wave phenomena revealed via a Hamiltonian eigenmode technique will be presented in Chapter 3. Inverse scattering transform (IST) technique, which can solve initial value problems of NLSE, KdV, and some other nonlinear partial differential equations (PDEs) will be introduced in Chapter 4.

2.1 Equations for Self-Consistent Description of Nonlinear Waves in Plasma

Plasma is a distinct state of matter containing electrically charged and neutral particles, including electrons, ions, and atoms/molecules, which enable plasma to conduct electricity and to react collectively to electromagnetic forces. Such unique properties separate plasma as the fourth state of matter, to distinguish it from the solid, liquid, and gas states. Plasma possesses nonlinear and dispersive properties; dissipation effect (collisions) can be neglected in many cases of study. The collective effect of charge particles enables plasma to support oscillations (waves) at high and low frequencies.

A uniform unmagnetized plasma is considered. Its fluid dynamic, with the aid of the ideal gas law, $P_a = n_a T_a$, where P, n, and T represent the pressure, density, and temperature, is defined by the

(1) continuity equations of the electron and ion fluids

$$\frac{\partial n_a}{\partial t} + \nabla \cdot n_a \mathbf{v_a} = 0 \tag{2.1}$$

and
(2) electron and ion momentum equations

$$n_a m_a \left(\frac{\partial \mathbf{v_a}}{\partial t} + \mathbf{v_a} \cdot \nabla \mathbf{v_a} \right) = -\nabla P_a + q_a n_a \mathbf{E} \tag{2.2}$$

where the subscript $a = e, i$, represents electron/ion fluid, respectively, $\nabla P_a = \gamma T_a \nabla n_a$, and $\gamma = C_p/C_v$ is the ratio of the specific heats C_p and C_v, at constant pressure and volume, respectively. In the adiabatic compression, $P/n^\gamma = \text{Const.}$, where $\gamma = (D+2)/D$, and D is the number of dimensions of the compression, i.e., $\gamma = 3, 2$, and $5/3$ for 1D, 2D, and 3D compression. In the isothermal case, $\gamma = 1$.

The electric fields in the momentum equations (2.2) are governed by Maxwell's equations

$$\nabla \times \mathbf{E} = -\frac{\partial \mathbf{B}}{\partial t} \tag{2.3a}$$

$$\nabla \times \mathbf{B} = \mu_0 \mathbf{J} + \frac{1}{c^2} \frac{\partial \mathbf{E}}{\partial t} \tag{2.3b}$$

$$\nabla \cdot \mathbf{E} = \frac{\rho}{\epsilon_0} \tag{2.3c}$$

$$\nabla \cdot \mathbf{B} = 0 \tag{2.3d}$$

where $\mathbf{J} = e(n_i\mathbf{v}_i - n_e\mathbf{v}_e)$ and $\rho = e(n_i - n_e)$ are the induced current density and charge density by electric fields $\mathbf{E}$ in plasma; singly charged ions are assumed; ϵ_0 is the free space permittivity. These two physical quantities ρ and $\mathbf{J}$ are related through the continuity (conservation of charge) equation

$$\frac{\partial \rho}{\partial t} + \nabla \cdot \mathbf{J} = 0 \tag{2.4}$$

The first two curl equations (2.3a) and (2.3b) are associated with Faraday's law and Ampere's Law and the next two divergence equations (2.3c) and (2.3d) are Gauss's law for the electric charges and magnetic charges.

2.2 Nonlinear Schrödinger Equation

2.2.1 *For electromagnetic wave*

Equations (2.3a) and (2.3b) are combined to become

$$\frac{\partial^2}{\partial t^2}\mathbf{E} - c^2\nabla^2\mathbf{E} = -\frac{1}{\epsilon_0}\frac{\partial}{\partial t}\mathbf{J}_e \tag{2.5a}$$

where $\mathbf{J}_e = \mathbf{J}_{eL} + \mathbf{J}_{eN} = -e(n_0 + \delta n)\mathbf{v}_{eL}$ is the (linear and nonlinear part of) induced electron current density by the EM wave field $\mathbf{E}$, $\mathbf{v}_{eL}$ is the electron linear velocity response to the EM wave field, and δn is the electron density perturbation induced by the radiation pressure of the wave modulation. With the aid of (2.2),

$$\frac{\partial}{\partial t}\mathbf{J}_{eL} = -en_0\frac{\partial}{\partial t}\mathbf{v}_{eL} = \epsilon_0\omega_p^2\mathbf{E} \tag{2.5b}$$

and

$$\delta n = -\frac{n_0}{2}\frac{\langle|\mathbf{v}_{eL}|\rangle^2}{\mathrm{v}_{te}^2}$$

where $\langle\ \rangle$ operates as a time average over high-frequency oscillation (a mode-type filter); the electron thermal speed $\mathrm{v}_{te} = \sqrt{T_e/m_e}$.

Consider a forward propagating wave modulation, which has a dominant carrier at (ω_0, k_0), a group velocity $v_g = k_0 c^2/\omega_0$, and a modulation amplitude $\psi(\mathbf{r}, t)$; the wave frequency ω_0 and propagation constant (wave number) k_0 are related by

$$\omega_0 = \sqrt{\omega_p^2 + k_0^2 c^2}$$

which is the linear dispersion relation of the electromagnetic wave propagating in an unmagnetized plasma having an electron plasma frequency of $\omega_p = \sqrt{\frac{n_0 e^2}{m_e \epsilon_0}}$. Substitute $E(\mathbf{r}, t) = \psi(\mathbf{r}, t) e^{i(\mathbf{k}_0 r - \omega_0 t)} +$ c.c. and $v_{eL} \sim -i\frac{e}{m_e \omega_0} \psi e^{i(\mathbf{k}_0 r - \omega_0 t)} +$ c.c. in (2.2) yields

$$\delta n \cong -n_0 \left(\frac{e}{m_e \omega_0}\right)^2 \frac{|\psi|^2}{v_{te}^2} \tag{2.5c}$$

Substitute (2.5b) and (2.5c) into (2.5a) and consider forward propagation in z (take forward scattering approximation to neglect $\partial^2/\partial t^2$ term), it simplifies (2.5a) to be

$$-i\left(\frac{\partial}{\partial t} + v_g \frac{\partial}{\partial z}\right)\psi(z, t) - \frac{1}{2}\frac{v_g}{k_0}\frac{\partial^2}{\partial z^2}\psi(z, t)$$

$$-\frac{\omega_p^2}{2\omega_0 v_{te}^2}\left(\frac{e}{m_e \omega_0}\right)^2 |\psi|^2 \psi(z, t) = 0 \tag{2.5d}$$

where the group velocity $v_g = \frac{k_0 c^2}{\omega_0}$.

Exercise 2.1: Derive the linear dispersion relation $\omega_0 = \sqrt{\omega_p^2 + k_0^2 c^2}$ of the electromagnetic wave propagating in an unmagnetized plasma.

Ans: Neglect δn in (2.5a) and find condition for non-trivial time-harmonic solution.

Exercise 2.2: Show that the forward scattering approximation leads (2.5a) to (2.5d).

Ans: $\frac{\partial^2}{\partial t^2}\psi(z, t)e^{i(k_0 z - \omega_0 t)} \sim -e^{i(k_0 z - \omega_0 t)}\left(2i\omega_0 \frac{\partial}{\partial t} + \omega_0^2\right)\psi(z, t)$ and $\frac{\partial}{\partial t}\delta n v_{eL} \sim \delta n \frac{\partial}{\partial t} v_{eL}$.

Introduce dimensionless coordinate and time, $\xi = k_0(z - v_g t) = k_0 z - \tau$ and $\tau = k_0 v_g t$, which convert the differential operators

$$\frac{\partial}{\partial t} = k_0 v_g \left(\frac{\partial}{\partial \tau} - \frac{\partial}{\partial \xi} \right), \quad \frac{\partial}{\partial z} = k_0 \frac{\partial}{\partial \xi}, \quad \text{and} \quad \frac{\partial^2}{\partial z^2} = k_0^2 \frac{\partial^2}{\partial \xi^2}$$

Then, in a moving frame at velocity $V = v_g$, (2.5d) is converted to a one-dimensional nonlinear Schrödinger equation it becomes (1.24)

$$-1/2 \frac{\partial^2}{\partial \xi^2} \varphi - \alpha |\varphi|^2 \varphi = i \frac{\partial}{\partial \tau} \varphi \qquad (2.5e)$$

where the normalized wave function $\varphi(\xi, \tau) = \frac{\psi(z,t)}{\psi_0}$, ψ_0 is the wave amplitude, and

$$\alpha = \frac{1}{2} \left(\frac{e}{m_e \omega_0 c} \right)^2 \left(\frac{\omega_p}{k_0 v_{te}} \right)^2 |\psi_0|^2$$

On the LHS of (2.5e), the first term induces wave dispersion and the second term steepens the wave during propagation.

Exercise 2.3: Show that the wave energy $\int |\varphi|^2 d\xi$ is conserved.

2.2.2 *For electron plasma (Langmuir) wave*

Electron plasma wave is an electrostatic wave, i.e., $\mathbf{E}_\ell = -\nabla \phi$, where ϕ is a scalar potential; thus, only (2.3c) in Maxwell's equations is involved in the formulation and is written explicitly as

$$\nabla \cdot \mathbf{E}_\ell = -\frac{e(n_e - \hat{n})}{\epsilon_0} = -\frac{e \delta n_{e\ell}}{\epsilon_0} \qquad (2.6a)$$

where $n_e = n_0 + \delta n_{e\ell} + n_s$; $\hat{n} = n_0 + n_s$; n_0, $\delta n_{e\ell}$, and n_s are the unperturbed plasma density and electron density perturbations associated with Langmuir waves and low-frequency oscillations, respectively. Apply the operator $\partial/\partial t$ to (2.1), i.e., $(\partial/\partial t)$ (2.1), and with the aid of (2.2) in which the convective term $(\mathbf{v}_e \cdot \nabla \mathbf{v}_e)$ is neglected, it yields

$$\frac{\partial^2}{\partial t^2}\delta n_{e\ell} - \mathrm{v}_{te}^2\nabla^2\delta n_{e\ell} = (e/m_e)[n_0\nabla\cdot\mathbf{E}_{\boldsymbol{\ell}} + \nabla\cdot(n_s\mathbf{E}_{\boldsymbol{\ell}})] \qquad (2.6b)$$

With the aid of (2.6a), a nonlinear mode equation governing the evolution of the Langmuir wave field $\mathbf{E}_\ell$ is derived to be

$$\left[\frac{\partial^2}{\partial t^2} + \omega_p^2 - 3v_{te}^2\nabla^2\right]\mathbf{E}_{\boldsymbol{\ell}} = -\omega_p^2\frac{n_s}{n_0}\mathbf{E}_{\boldsymbol{\ell}} \qquad (2.6c)$$

The nonlinear nature of the equation is shown implicitly by the RHS term of (2.6c), in which the wave-induced background density perturbation, n_s, modifies the dispersion property of the Langmuir wave self-consistently. The governing equation of this background density perturbation is derived in the following.

The momentum equations of electron and ion fluids are combined by adding them together. The electric field terms in the two equations cancel each other; and the electron inertial term $m_e\partial\mathbf{v}_e/\partial t$ and the ion convective term $m_i\mathbf{v}_i\cdot\nabla\mathbf{v}_i$, which are small compared to their respective counterpart, are neglected. The combined equation is obtained to be

$$m_i\frac{\partial}{\partial t}\mathbf{v}_i + m_e\mathbf{v}_e\cdot\nabla\mathbf{v}_e = -\frac{1}{n_0}\nabla(P_e + P_i) \qquad (2.6d)$$

Apply the operator $(\nabla\cdot)$ to (2.6d), i.e., $(\nabla\cdot)$ (2.6d), and with the aid of the quasi-neutrality $n_{se} = n_{si} = n_s$, and the relations $\mathbf{v}_e\cdot\nabla\mathbf{v}_e = \nabla(v_e^2/2)$ and $\partial n_s/\partial t + \nabla\cdot(n_0\mathbf{v}_i) = 0$ of the continuity equation, (2.6d) becomes

$$\left[\frac{\partial^2}{\partial t^2} - C_s^2\nabla^2\right]\left(\frac{n_s}{n_0}\right) = \frac{m_e}{m_i}\nabla^2\left\langle\frac{v_e^2}{2}\right\rangle \qquad (2.6e)$$

where $C_s = [(T_e + 3T_i)/m_i]^{1/2}$ is the ion acoustic speed; the implicit nonlinear term on the RHS of (2.6e) gives rise to a ponderomotive force acting on the electron plasma.

This force results from an average effect of the non-uniform electron quiver motion in the Langmuir wave fields; as the electron plasma is pushed by this force, the induced self-consistent electric field pulls ion plasma to move together to setup background density perturbation n_s as illustrated in Fig. 2.1.

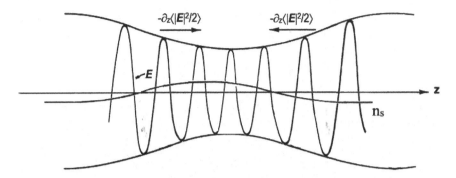

Fig. 2.1. Density perturbation n_s set up by ponderomotive forces.

This density perturbation varies slowly in time so that the time derivative term on the LHS of (2.6e) can be neglected. Thus, (2.6e) leads to

$$\frac{n_s}{n_0} \sim -\frac{m_e}{m_i} \left\langle \frac{v_{e\ell}^2}{2C_s^2} \right\rangle \tag{2.6f}$$

Again, consider a modulated wave forward propagating in z; it has a dominant carrier at (ω_1, k_1), a group velocity $v_{g\ell} = 3k_1 v_{te}^2/\omega_1$, and a modulation amplitude $\psi_\ell(z,t)$, where

$$\omega_1 = \sqrt{\omega_p^2 + 3k_1^2 v_{te}^2}$$

is the linear dispersion relation of the Langmuir wave. Substitute $E_\ell(z,t) = \psi_\ell(z,t)e^{i(k_1 z - \omega_1 t)} +$ c.c. in (2.2) obtains

$$v_{e\ell} \sim -i\frac{e}{m_e \omega_1}\psi_\ell e^{i(k_1 z - \omega_1 t)} + \text{c.c.}$$

which is then substituted into (2.6f) to give

$$\frac{n_s}{n_0} \sim -\frac{m_e}{m_i}\left(\frac{e}{m_e \omega_1}\right)^2 \frac{|\psi_\ell|^2}{C_s^2} \tag{2.6g}$$

Substitute $E_\ell(z,t) = \psi_\ell(z,t)e^{i(k_1 z - \omega_1 t)} +$ c.c. into (2.6c), apply the same forward scattering approximation employed in Section 2.2.1, introduce dimensionless coordinate and time, $\xi_1 = k_1(z - v_{g\ell}t) =$

$k_1 z - \tau_1$ and $\tau_1 = k_1 v_{g\ell} t$, in a moving frame at velocity $V = v_{g\ell}$, and normalize the wave function:

$$\varphi_\ell(\xi_1, \tau_1) = \frac{\psi_\ell(z, t)}{\psi_{\ell 0}}$$

(2.6c) is converted to a similar (to (2.5e)) one-dimensional nonlinear Schrödinger equation

$$-\frac{1}{2}\frac{\partial^2}{\partial \xi_1^2}\varphi_\ell - \alpha_1 |\varphi_\ell|^2 \varphi_\ell = i\frac{\partial}{\partial \tau_1}\varphi_\ell \qquad (2.7)$$

where

$$\alpha_1 = \frac{1}{6}\left(\frac{e}{m_e \omega_1}\right)^2 \left(\frac{\omega_{pi}}{k_1 v_{te} C_s}\right)^2 |\psi_{\ell 0}|^2$$

and $\omega_{pi} = \sqrt{\frac{n_0 e^2}{m_i \epsilon_0}}$ is the ion plasma frequency.

Exercise 2.4: Derive the linear dispersion relation $\omega_1 = \sqrt{\omega_p^2 + 3k_1^2 v_{te}^2}$ of the Langmuir wave.

Ans: Neglect n_s in (2.6c) and find condition for non-trivial time-harmonic solution.

2.3 Korteweg–de Vries (KdV) Equation for Ion Acoustic Wave

The propagation of ion waves (along z axis) is formulated through (2.1) and (2.2). Without the assumption of quasi-neutrality, (2.3c) is included to relate the wave field to the density oscillations. Add the electron and ion momentum equations and neglect the electron inertial and convective terms, a one-fluid momentum equation is obtained to be

$$\left[\frac{\partial}{\partial t}V_{si} + \frac{\partial}{\partial z}\left(\frac{V_{si}^2}{2}\right)\right] = -C_s^2\frac{\partial}{\partial z}\left(\frac{\delta n_{si}}{n_0}\right) - \left(\frac{T_e}{m_i}\right)\frac{\partial}{\partial z}\left(\frac{\delta n_{se} - \delta n_{si}}{n_0}\right)$$
$$(2.8a)$$

where δn_{se} and δn_{si} are the electron and ion density perturbation and V_{si} is the velocity perturbation of the ion fluid in the presence of an ion wave. From the electron momentum equation (with the inertial terms neglected), it gives

$$E_s \sim -\frac{T_e}{n_0 e}\frac{\partial}{\partial z}\delta n_{se} \sim -\frac{T_e}{n_0 e}\frac{\partial}{\partial z}\delta n_{si}$$

where E_s is the self-consistent field induced in the ion wave to keep ambipolar motion (i.e., $V_{se} \sim V_{si}$). With the aid of Gauss's law (2.3c):

$$(\delta n_{se} - \delta n_{si}) = -\frac{\epsilon_0}{e}\frac{\partial}{\partial z}E_s \sim \frac{v_s}{\omega_{pi}}\frac{\partial^2}{\partial z^2}\delta n_{si}$$

Equation (2.8a) becomes

$$\left[\frac{\partial}{\partial t}V_{si} + \frac{\partial}{\partial z}\left(\frac{V_{si}^2}{2}\right)\right] = C_s^2\frac{\partial}{\partial z}\left(\frac{\delta n_{si}}{n_0}\right) - \left(\frac{v_s^4}{\omega_{pi}^2}\right)\frac{\partial^3}{\partial z^3}\left(\frac{\delta n_{si}}{n_0}\right) \quad (2.8b)$$

where $v_s^2 = \frac{T_e}{m_i}$, and $\frac{\omega_{pi}^2}{v_s^2} = k_{De}^2$, square of the electron Debye number. Take the partial time derivative operation $\frac{\partial}{\partial t}$ on both sides of (2.8b) and the aid of the ion continuity equation $\frac{\partial}{\partial t}\frac{\delta n_{si}}{n_0} = -\frac{\partial}{\partial z}V_{si}$, (2.8b) becomes

$$\frac{\partial}{\partial t}\left[\frac{\partial}{\partial t}V_{si} + \frac{\partial}{\partial z}\left(\frac{V_{si}^2}{2}\right)\right] = C_s^2\frac{\partial^2}{\partial z^2}V_{si} + \left(\frac{v_s^4}{\omega_{pi}^2}\right)\frac{\partial^4}{\partial z^4}V_{si} \quad (2.8c)$$

In the linear regime, the second (nonlinear) term in the LHS bracket and the second (dispersion) term on the RHS of (2.8c) are negligible; then, (2.8c) reduces to $(\frac{\partial^2}{\partial t^2} - C_s^2\frac{\partial^2}{\partial z^2})V_{si} = 0$, which defines the linear dispersion relation $\omega_s = kC_s$ for the ion acoustic wave.

Equation (2.8c) is transformed to a moving frame at the ion acoustic velocity C_s by setting

$$t_1 = t \text{ and } z_1 = z - C_s t$$

and letting $V_{si}(z,t) = A_s(z_1, t_1)$; the differential operators convert to

$$\frac{\partial}{\partial t} \rightarrow \frac{\partial}{\partial t_1} - C_s\frac{\partial}{\partial z_1} \text{ and } \frac{\partial}{\partial z} \rightarrow \frac{\partial}{\partial z_1}$$

and (2.8c) becomes

$$\frac{\partial}{\partial t_1}\left[\frac{\partial}{\partial t_1}A_s + \frac{\partial}{\partial z_1}\left(\frac{A_s^2}{2}\right)\right] - 2C_s\frac{\partial}{\partial z_1}\left[\frac{\partial}{\partial t_1}A_s + \frac{\partial}{\partial z_1}\left(\frac{A_s^2}{4}\right)\right]$$

$$= \left(\frac{v_s^4}{\omega_{pi}^2}\right)\frac{\partial^4}{\partial z_1^4}A_s \tag{2.8d}$$

In the moving frame, the frequencies of the linear ion acoustic waves are downshifted to zero, thus, $\left|\frac{\partial}{\partial t_1}\right| \ll \left|C_s\frac{\partial}{\partial z_1}\right|$, and the first term on the LHS of (2.8d) is neglected. Introduce dimensionless variables and function:

$$\tau = \omega_{pi}t_1, \quad \eta = \beta z_1, \quad \text{and} \quad \phi(\eta, \tau) = \alpha_2 A_s(z_1, t_1)$$

where

$$\beta = \left(\frac{2C_s\omega_{pi}^3}{v_s^4}\right)^{1/3} = \left(\frac{2C_s}{v_s}\right)^{1/3} k_{De} \quad \text{and}$$

$$\alpha_2 = \frac{\beta}{12\omega_{pi}} = \frac{1}{12v_s}\left(\frac{2C_s}{v_s}\right)^{1/3}$$

Equation (2.8d) is converted to a standard KdV equation (1.1)

$$\frac{\partial}{\partial \tau}\phi + \frac{\partial^3}{\partial \eta^3}\phi + 6\phi\frac{\partial}{\partial \eta}\phi = 0 \tag{2.8e}$$

The first term characterizes the time evolution rate of the wave propagating under the influence of dispersion effect (second term) and the nonlinear steepening effect (third term) of the medium.

2.4 Burgers Equation for Dissipated Ion Acoustic Wave

In collision plasma, collision terms are added to (2.2); one in the form of $-n_a m_a \nu_{ab}(v_a - v_b)$ is ascribed to collisions between electrons and ions, and the other one in the form of $-n_a m_a \nu_a v_a$ is ascribed to collisions with each other among the same species and with the background neutral particles. Conservation of momentum imposes that

$n_a m_a \nu_{ab} = n_b m_b \nu_{ba}$; thus, in the one-fluid equation, the terms associated with electron–ion collisions cancel each other. Quasi-neutrality is assumed, i.e., $\delta n_{se} = \delta n_{si}$, and the remaining collision terms represent the net collision damping effect on the ion wave; in fact, such collisions cause diffusion of the perturbations, which in turn, causes ion wave to dampen. Thus, the remaining collision terms in the one-fluid equation are modeled by a single diffusion term in the form of $\frac{\partial^2}{\partial z^2} V_{si}$, where the diffusion coefficient $D = \frac{\nu_i v_{ti}^2}{\omega_{pi}^2}$, and (2.8a) is modified to be

$$\left[\frac{\partial}{\partial t} V_{si} + \frac{\partial}{\partial z} \left(\frac{V_{si}^2}{2} \right) \right] = -C_s^2 \frac{\partial}{\partial z} \left(\frac{\delta n_{si}}{n_0} \right) + D \frac{\partial^2}{\partial z^2} V_{si} \qquad (2.9a)$$

Next, take the partial time derivative operation $\frac{\partial}{\partial t}$ on both sides of (2.9a) and apply the ion continuity equation $\frac{\partial}{\partial t} \frac{\delta n_{si}}{n_0} = -\frac{\partial}{\partial z} V_{si}$, it results to

$$\frac{\partial}{\partial t} \left[\frac{\partial}{\partial t} V_{si} + \frac{\partial}{\partial z} \left(\frac{V_{si}^2}{2} \right) \right] = C_s^2 \frac{\partial^2}{\partial z^2} V_{si} + D \frac{\partial}{\partial t} \frac{\partial^2}{\partial z^2} V_{si} \qquad (2.9b)$$

Again, like (2.8c), (2.9b) is transformed to a moving frame at the ion acoustic velocity C_s by setting

$$t_1 = t \text{ and } z_1 = z - C_s t$$

and letting $V_{si}(z,t) = A_{s1}(z_1, t_1)$; again,

$$\frac{\partial}{\partial t} \to \frac{\partial}{\partial t_1} - C_s \frac{\partial}{\partial z_1} \text{ and } \frac{\partial}{\partial z} \to \frac{\partial}{\partial z_1}$$

Then, (2.9b) becomes

$$\frac{\partial}{\partial t_1} \left[\frac{\partial}{\partial t_1} A_{s1} + \frac{\partial}{\partial z_1} \left(\frac{A_{s1}^2}{2} \right) - D \frac{\partial^2}{\partial z_1^2} A_{s1} \right]$$

$$-2C_s \frac{\partial}{\partial z_1} \left[\frac{\partial}{\partial t_1} A_{s1} + \frac{\partial}{\partial z_1} \left(\frac{A_{s1}^2}{4} \right) \right]$$

$$= -C_z D \frac{\partial^3}{\partial z_1^3} A_{s1} \qquad (2.9c)$$

In the moving frame, $|\frac{\partial}{\partial t_1}| \ll |C_s \frac{\partial}{\partial z_1}|$, the first term on the LHS of (2.9c) is negligible. Introduce the similar dimensionless variables as those introduced in the KdV case and a normalized function:

$$\tau = \omega_{pi} t_1, \ \eta = \beta z_1, \ \text{and} \ \phi_1(\eta, \tau) = \alpha_3 A_{s1}(z_1, t_1)$$

where

$$\beta = \left(\frac{2C_s \omega_{pi}^3}{v_s^4} \right)^{\frac{1}{3}} = \left(\frac{2C_s}{v_s} \right)^{\frac{1}{3}} k_{De}$$

$$\text{and} \ \alpha_3 = \frac{\beta}{2\omega_{pi}} = \frac{1}{2v_s} \left(\frac{2C_s}{v_s} \right)^{1/3} = 6\alpha_2$$

Equation (2.9c) is converted to a standard Burgers equation

$$\frac{\partial}{\partial \tau} \phi_1 + \phi_1 \frac{\partial}{\partial \eta} \phi_1 = b \frac{\partial^2}{\partial \eta^2} \phi_1 \tag{2.10}$$

where $b = \frac{D}{2\omega_{pi}} \beta^2$. The diffusion term on the RHS of (2.10) introduces damping (flattening) effect and the nonlinear convection term on the LHS (second term) steepens the wave.

Exercise 2.5: In the case of $b = 0$, (2.10) can be solved analytically by the method of characteristics to obtain an implicit solution. For an initial condition, $\phi_1(\eta, 0) = \psi_0(\eta_0)$, where η_0 is the initial position of a characteristic, find the implicit solution.

Ans: $\phi_1(\eta, \tau) = \psi_0(\eta - \phi_1(\eta, \tau)\tau)$.

2.5 Upper Hybrid Soliton Generated in Ionospheric HF Heating Experiments

The F-region of the ionosphere has been a platform for experimental and theoretical investigation of nonlinear wave generation in magnetized plasma by ground-transmitted powerful HF waves. In experiments, the O-mode HF heating waves pass the upper hybrid resonance layer before reaching the reflection height in an over-dense ionosphere (i.e., the HF wave frequency f_0 is less than the maximum electron plasma frequency foF2); upper hybrid waves are excited parametrically by the HF heaters; those waves form standing waves across the magnetic field $\mathbf{B}_0 = \hat{z} B_0$ and propagate along the magnetic field ($\pm \hat{z}$ direction).

2.5.1 *Plasma density perturbed by the parametrically excited upper hybrid waves*

Ponderomotive forces induced by parametrically excited upper hybrid waves move electrons in the magnetic field direction and ions follow the electron motion via the induced ambipolar electric field. As a result, plasma density variations along the magnetic field are formed.

The governing equation of plasma density variation $\delta n(z) = \delta n_e = \delta n_i$, driven by the axial ponderomotive forces induced by the upper hybrid wave pressure is derived as follows.

The continuity and momentum equations (2.1) and (2.2) for perturbation physical quantities become

$$\partial_t \, \delta n_e + n_0 \nabla \cdot \delta \mathbf{v_e} = 0 = \partial_t \, \delta n_i + n_0 \nabla \cdot \delta \mathbf{v_i} \qquad (2.11)$$

$$m_e(\partial_t \, \delta \mathbf{v_e} + \langle \mathbf{v_e} \cdot \nabla \mathbf{v_e} \rangle) = -m_e v_{te}^2 \nabla_z \frac{\delta n_e}{n_0} - e\delta \mathbf{E} \qquad (2.12)$$

$$m_i(\partial_t \, \delta \mathbf{v_i} + \langle \mathbf{v_i} \cdot \nabla \mathbf{v_i} \rangle) = -3m_i v_{ti}^2 \nabla_z \frac{\delta n_i}{n_0} + e\delta \mathbf{E} \qquad (2.13)$$

where $v_{te,i} = (T_{e,i}/m_{e,i})^{1/2}$ and $\langle \, \rangle$ operates as a mode-type filter.

With the aid of (2.11), (2.12) and (2.13) are combined to be

$$(\partial_t^2 - C_s^2 \nabla_z^2)\frac{\delta n}{n_0} = (m_e/m_i)\nabla_z^2 \left\langle \frac{v_e^2}{2} \right\rangle \qquad (2.14a)$$

where the electron inertial term is neglected; $C_s = [(T_e + 3T_i)/m_i]^{1/2}$ is the ion acoustic speed and m_i is the ion (O^+) mass; $m_i|\langle \mathbf{v_i} \cdot \nabla \mathbf{v_i} \rangle| \ll m_e|\langle \mathbf{v_e} \cdot \nabla \mathbf{v_e} \rangle|$ is applied.

Since δn_e is a non-oscillatory density perturbation, $|\partial_t^2(\delta n/n_0)| \ll |C_s^2 \nabla_z^2(\delta n/n_0)|$; then, (2.14a) is approximated to obtain

$$\frac{\delta n}{n_0} \cong -\frac{m_e}{m_i} \frac{\langle V_{ue}^2 \rangle}{2C_s^2} \qquad (2.14b)$$

where V_{ue} is the electron quiver velocity in the upper hybrid wave fields.

2.5.2   *Nonlinear envelope equation of the upper hybrid waves*

Only electrons respond effectively to the upper hybrid wave field, the formulation involves electron fluid equations (2.1) and (2.2), which can be combined into an equation for the upper hybrid wave electrostatic potential ϕ_u, given by the upper hybrid wave field $\mathbf{E}_u = -\nabla\phi_u$.

The continuity and momentum equations (adding Lorentz force in (2.2)) and Poisson's equation, for the upper hybrid wave density, velocity, and electrostatic potential perturbations, n_{ue}, $\mathbf{v}_{ue}$, and ϕ_u, in magnetized plasma, are given as

$$\partial_t\, n_{ue} + \langle \nabla \cdot (n_0 + \delta n)\delta\mathbf{v}_{ue}\rangle = 0 \qquad (2.15)$$

$$\partial_t\, \mathbf{v}_{ue} - \Omega_e\, \mathbf{v}_{ue} \times \hat{z} = -3v_{te}^2 \nabla_z \frac{n_{ue}}{n_0} + \frac{e}{m_e}\nabla\phi_u \qquad (2.16)$$

$$\nabla^2\phi_u = \frac{en_{ue}}{\epsilon_0} \qquad (2.17)$$

where $\Omega_e = eB_0/m_e$ is the electron cyclotron frequency, the convective term in (2.16) is separated to include in (2.12), and isotropic pressure term on the right-hand side (RHS) of (2.16) is assumed in the case of ionospheric plasma, where the electron cyclotron frequency is much smaller than the electron plasma frequency.

Equations (2.15)–(2.17) are then combined into a nonlinear equation for the upper hybrid wave potential ϕ_u as

$$[(\partial_t^2 + \Omega_e^2)(\partial_t^2 + \omega_p^2 - 3v_{te}^2\nabla^2)\nabla^2 - \Omega_e^2(\omega_p^2 - 3v_{te}^2\nabla^2)\nabla_\perp^2]\phi_u$$
$$\cong -(\omega_p^2 - 3v_{te}^2\nabla^2)(\partial_t^2\nabla^2 + \Omega_e^2\nabla_z^2)\phi_u\frac{\delta n}{n_0} \qquad (2.18)$$

Exercise 2.6: Determine the linear dispersion relation of the upper hybrid mode.

Ans: $\omega = (\omega_{UH}^2 + 3k^2v_{te}^2)^{1/2}$, where $\omega_{UH} = (\omega_p^2 + \Omega_e^2)^{1/2}$ is the upper hybrid resonance frequency.

In the following, (2.18) is analyzed with forward scattering approximation. Set $\phi_u = \varphi(t, z)e^{-i\omega_u t}\cos k_u x\, e^{ik_z z} + $ c.c., where $\omega_u = (\omega_p^2 + \Omega_e^2 + 3k^2v_{te}^2)^{1/2} > \Omega_e$, $k^2 = k_u^2 + k_z^2$, and $|k_u| \gg |k_z|$; from the electron momentum equation (2.16), the electron quiver velocity in the upper hybrid wave fields is obtained to be

$$V_{uex} \sim -i\left(\frac{\omega_u}{\omega_u^2 - \Omega_e^2}\right)\left[\frac{ek_u\varphi(t,z)}{m_e}\right]e^{-i\omega_u t}\sin k_u x e^{ik_z z} + \text{c.c.}$$

and

$$V_{uey} \sim -\left(\frac{\Omega_e}{\omega_u^2 - \Omega_e^2}\right)\left[\frac{ek_u\varphi(t,z)}{m_e}\right]e^{-i\omega_u t}\sin k_u x e^{ik_z z} + \text{c.c.}$$

which lead to

$$\langle V_{ue}^2\rangle = \langle V_{uex}^2 + V_{uey}^2\rangle \cong \left[\frac{\omega_u^2 + \Omega_e^2}{(\omega_u^2 - \Omega_e^2)^2}\right]\left[\frac{ek_u|\varphi(t,z)|}{m_e}\right]^2$$

Since $|\partial_t| > \Omega_e$ and $|\nabla_\perp| \gg |\nabla_z|$, and set $V_g = 2k_z A$, $A = 3v_{te}^2/2\omega_u$ and $B = (e^2 k_u^2/4m_e m_i \omega_u C_s^2)[(\omega_u^2 + \Omega_e^2)/(\omega_u^2 - \Omega_e^2)]$, (2.18) reduces to

$$-i(\partial_t + V_g\partial_z)\varphi - A\partial_z^2\varphi - B|\varphi|^2\varphi = 0 \qquad (2.19)$$

In a moving frame, $\eta = \sqrt{\frac{1}{A}}(z - V_g t)$ and $\tau = t$, and set $\phi = \sqrt{\frac{2}{B}}\varphi$; (2.19) reduces to the cubic nonlinear Schrödinger equation (1.25a).

The theory and analysis show that the HF heater excited localized upper hybrid waves (i.e., $k_z \cong 0$) can form a solitary wave in the upper hybrid resonance region, where the background plasma density is pushed out by the upper hybrid soliton to form a caviton. A remote sensor called "ionosonde" (an HF radar) is usually applied to monitor the plasma density profile on the bottom side of the ionosphere, via recorded ionograms; caviton induces a bump, around the upper hybrid resonance frequency, in the virtual height spread of the ionogram trace. This is exemplified in Figs. 2.2(a) and 2.2(b), which are the ionograms, recorded from the results of the experiments conducted using the HAARP transmitter facility at Gakona, AK, at full power (3.6 MW), with the HF heater transmitting at 3.2 MHz directed along the geomagnetic zenith. As shown in Fig. 2.2(a), with the O-mode heater turned on for 2 minutes, significant virtual height spread was observed in this heater-off ionogram. Moreover, there is a noticeable bump in the virtual height spread of the ionogram trace.

This bump appears adjacent to the plasma frequency ($f_p = (f_0^2 - f_{ce}^2)^{1/2} \sim 2.88$ MHz) of the upper hybrid resonance layer of

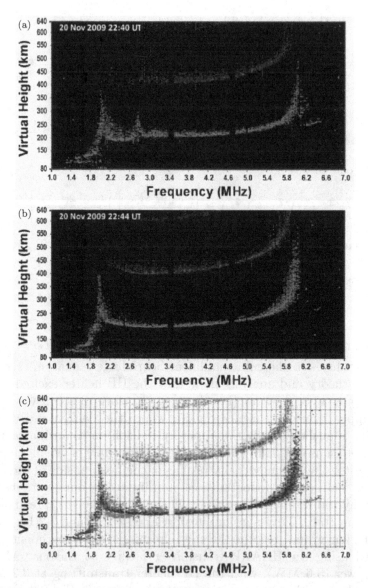

Fig. 2.2.   Two contrasting ionograms show the impact of the O-mode HF heater
on the ionosphere. (a) Recorded at the moment the O mode heater turns off after
being on for 2 minutes, (b) recorded at 4 minutes later, which is considered to be
an ionogram of the recovered background, and (c) superposed ionogram.

the ($f_0 = 3.2$ MHz) O-mode heater, where $f_{ce} \sim 1.4$ MHz is the electron cyclotron frequency. This bump manifests a self-induced plasma density cavity which traps an upper hybrid soliton to balance plasma pressure.

Problems

P2.1. In a linear system with a prescribed potential distribution $V(\xi)$, the linear Schrödinger equation is

$$-1/2\frac{\partial^2}{\partial\xi^2}\varphi + V(\xi)\varphi = i\frac{\partial}{\partial\tau}\varphi \qquad (P2.1)$$

If $V(\xi)$ is a localized potential well, then there will be a finite number of bound states with discrete eigenenergies $E_n = -\alpha_n^2$, $n = 1, 2, \ldots, N$ and a continuum of states with eigenenergies $E = k^2$. These eigenenergies determine the corresponding eigenfunctions asymptotically (i.e., functions in the region $\xi \to \pm\infty$, where $V \to 0$). Find the asymptotic $\varphi(\xi, \tau)$ of the bound and unbound states.

P2.2. Show that the nonlinear Schrödinger equation (2.7) indicates that $\int_{-\infty}^{\infty} |\varphi_\ell|^2 d\xi_1$ is conserved.

P2.3. Galilean invariance: if $\phi(\eta, \tau)$ is a solution of the KdV equation (2.8e), show that $\tilde{\phi}(\eta, \tau) = \phi(\eta - 6v\tau, \tau) + v$ is also a solution.

P2.4. Burgers equation (2.10) has the initial condition: $\phi_1(\eta, 0) = a\eta + b$, find an explicit solution $\phi_1(\eta, \tau)$ to (2.10).

Chapter 3

Characteristics of Nonlinear Waves

In Chapter 2, nonlinear equations descriptive of wave propagation in plasma are formulated. Those include the nonlinear Schrödinger equation (NLSE), Korteweg–de Vries (KdV) equation, and Burgers equation, which are generic equations also applicable for the description of similar nonlinear wave phenomena in other media. In the following, noticeable characteristic features of the solutions of these nonlinear equations are presented. Hamiltonian eigenmode technique is applied to reveal nonlinear wave phenomena and to determine appropriate "mode" solutions of the NLSE and KdV equations. Cole–Hopf transform is applied to solve the Burgers equation.

3.1 Nonlinear Schrödinger Equation (NLSE)

Most wave propagation in weakly nonlinear, dispersive, energy-preserving systems is canonically descriptive, in an appropriate limit, by the NLSE. Specifically, the NLSE describes the evolution of a wave packet in a weakly nonlinear and dispersive medium when dissipation can be neglected. It has been applied for the study of optical pulse propagation in nonlinear fibers, the phenomenon of self-focusing, and the conditions under which an electromagnetic beam can propagate without spreading in nonlinear dispersive media.

3.1.1   *Characteristic features of solutions*

It is hard to find a general analytical solution of the NLSE (1.24), however, it will be a good idea to know what are the rigid constraints (attributes) that the equation imposes on its solutions. Those involve conservation laws, symmetries, and invariances.

3.1.1.1   *Conservation laws*

The conservation laws are in the form of the continuity equation $\frac{\partial}{\partial \tau} N + \frac{\partial}{\partial \xi} J = 0$, where N is the density of a conserved quantity and J is the flux of this quantity. Based on (1.24), the first three conservation laws are illustrated in the following:

$$\frac{\partial}{\partial \tau} |\varphi|^2 + \frac{\partial}{\partial \xi} \left[\frac{i}{2} |\varphi|^2 \frac{\partial}{\partial \xi} \left(\ln \frac{\varphi^*}{\varphi} \right) \right] = 0 \qquad (3.1a)$$

$$\frac{\partial}{\partial \tau} \left[\frac{i}{2} |\varphi|^2 \frac{\partial}{\partial \xi} \left(\ln \frac{\varphi^*}{\varphi} \right) \right]$$
$$+ \frac{\partial}{\partial \xi} \left(\left| \frac{\partial \varphi}{\partial \xi} \right|^2 - \frac{1}{2} \alpha |\varphi|^4 - \frac{1}{4} \frac{\partial^2}{\partial \xi^2} |\varphi|^2 \right) = 0 \qquad (3.1b)$$

$$\frac{\partial}{\partial \tau} \left(\left| \frac{\partial \varphi}{\partial \xi} \right|^2 - \alpha |\varphi|^4 \right)$$
$$+ \frac{\partial}{\partial \xi} \left\{ -\frac{i}{2} \left[\left(\frac{\partial \varphi^*}{\partial \xi} \frac{\partial^2}{\partial \xi^2} \varphi - \frac{\partial \varphi}{\partial \xi} \frac{\partial^2}{\partial \xi^2} \varphi^* \right) \right. \right.$$
$$\left. \left. + 2\alpha |\varphi|^2 \left(\varphi \frac{\partial \varphi^*}{\partial \xi} - \varphi^* \frac{\partial \varphi}{\partial \xi} \right) \right] \right\} = 0 \qquad (3.1c)$$

The physical quantities are identified as

$$N_1 = |\varphi|^2 \text{ and } J_1 = \frac{i}{2} |\varphi|^2 \frac{\partial}{\partial \xi} \left(\ln \frac{\varphi^*}{\varphi} \right) = N_2$$

$$J_2 = \left| \frac{\partial \varphi}{\partial \xi} \right|^2 - \frac{1}{2} \alpha |\varphi|^4 - \frac{1}{4} \frac{\partial^2}{\partial \xi^2} |\varphi|^2$$

$$= \frac{1}{2} \left(\left| \frac{\partial \varphi}{\partial \xi} \right|^2 - \alpha |\varphi|^4 \right) - \frac{1}{4} \left(\varphi^* \frac{\partial^2}{\partial \xi^2} \varphi + \varphi \frac{\partial^2}{\partial \xi^2} \varphi^* \right)$$

$$N_3 = \left| \frac{\partial \varphi}{\partial \xi} \right|^2 - \alpha \, |\varphi|^4 \text{ and}$$

$$J_3 = -\frac{i}{2} \left[\left(\frac{\partial \varphi^*}{\partial \xi} \frac{\partial^2}{\partial \xi^2} \varphi - \frac{\partial \varphi}{\partial \xi} \frac{\partial^2}{\partial \xi^2} \varphi^* \right) + 2\alpha \, |\varphi|^2 \left(\varphi \frac{\partial \varphi^*}{\partial \xi} - \varphi^* \frac{\partial \varphi}{\partial \xi} \right) \right]$$

$$= \text{Im} \left(\frac{\partial \varphi^*}{\partial \xi} \frac{\partial^2}{\partial \xi^2} \varphi + 2\alpha \, |\varphi|^2 \, \varphi \frac{\partial \varphi^*}{\partial \xi} \right)$$

where $N_1, J_1 = N_2, J_2, N_3$, and J_3 represent the mass density, momentum density, pressure, energy density, and energy flux, respectively, of a unit mass particle system.

In addition, the Hamiltonian of the system is defined as

$$H = 1/2 \int_{-\infty}^{\infty} \left(\left| \frac{\partial \varphi}{\partial \xi} \right|^2 - \alpha \, |\varphi|^4 \right) d\xi \tag{3.1d}$$

The conservation law (3.1c) implies that $\frac{dH}{d\tau} = 0$; thus, H is a constant of motion and is given by the initial value

$$H = 1/2 \int_{-\infty}^{\infty} \left(\left| \frac{\partial \varphi_0}{\partial \xi} \right|^2 - \alpha \, |\varphi_0|^4 \right) d\xi$$

Exercise 3.1: Verify the conservation laws (3.1) to (3.1c).

Exercise 3.2: Equation (1.32a) is a solution of (1.24) with $t = \frac{\tau}{2}$, $x = \xi$, and $\alpha = 1$; apply this solution to find the Hamiltonian H of the system.

Ans: $H = \frac{16}{3} \kappa^4$.

3.1.1.2 *Scaling symmetry*

If $\varphi(\xi, \tau)$ is a solution of (1.24) with the initial condition $\varphi(\xi, 0) = \varphi_0(\xi)$, then

$$\varphi_\alpha(\xi, \tau) = \alpha^{-1} \varphi \left(\frac{\xi}{\alpha}, \frac{\tau}{\alpha^2} \right) \tag{3.1e}$$

is also a solution of (1.24) with the initial condition $\varphi_\alpha(\xi, 0) = \alpha^{-1} \varphi \left(\frac{\xi}{\alpha}, 0 \right) = \alpha^{-1} \varphi_0 \left(\frac{\xi}{\alpha} \right)$.

Exercise 3.3: Verification of the scaling relation (3.1e) of the solution of the NLSE (1.24).

3.1.1.3 *Galilean invariance*

If $\varphi(\xi, \tau)$ is a solution of (1.24) with the initial condition $\varphi(\xi, 0) = \varphi_0(\xi)$, then

$$\varphi_k(\xi, \tau) = e^{ik\xi} e^{\frac{i|k|^2 \tau}{2}} \varphi(\xi - k\tau, \tau) \tag{3.1f}$$

is also a solution of (1.24) with the initial condition $\varphi_k(\xi, 0) = e^{ik\xi} \varphi(\xi, 0) = e^{ik\xi} \varphi_0(\xi)$.

Exercise 3.4: Apply (3.1f) to (1.32b) and show that φ_k is a solution of (1.25a).

3.1.1.4 *Virial theorem (variance identity)*

A variance $V(\tau)$ is defined as

$$V(\tau) = \int_{-\infty}^{\infty} \xi^2 |\varphi(\xi, \tau)|^2 d\xi \tag{3.1g}$$

Apply (1.24) and integration by parts, the variance identity is obtained as

$$\frac{d^2}{d\tau^2} V(\tau) = 4H + \alpha \int |\varphi|^4 \, d\xi \tag{3.1h}$$

It is noted that $V(\tau) \geq 0$ for all τ. This equation will be generalized to the multidimensional cases in Section 3.1.3 to consider wave collapse.

Exercise 3.5: Derive the variance identity (3.1h).

Ans: Convert time derivative to spatial derivative via (1.24).

3.1.2 *Eigen solutions*

The nonlinear Schrödinger equation (1.24) is analyzed by first converting it to an eigenvalue equation. It is done by introducing

$\varphi(\xi, \tau) = \phi(\xi)e^{-iE\tau}$, where E is the eigenvalue of a state and $\phi(\xi)$ is a real function. An eigenvalue equation is obtained as

$$\left(-1/2\frac{d^2}{d\xi^2} - \alpha\phi^2\right)\phi = E\phi \qquad (3.2)$$

Consider (ϕ, ξ) as the equivalent spatial coordinate and time of a system, (3.2) represents an equation of motion of a unit mass object moving in this one-dimensional space and interacting with a restoring force of $-2(E\phi + \alpha\phi^3)$ (a nonlinear oscillator model). Multiply $\frac{d}{d\xi}\phi$ to both sides of (3.2), it leads to

$$\frac{d}{d\xi}\left[1/2\left(\frac{d\phi}{d\xi}\right)^2 + \left(E\phi^2 + 1/2\alpha\phi^4\right)\right] = 0 \qquad (3.3)$$

It is recognized that the quantity in the parentheses on the LHS of (3.3) is invariant to the equivalent time "ξ". Hence, this unit mass object moving in a potential field $V(\phi) = E\phi^2 + 1/2\alpha\phi^4$ with a kinetic energy $T = 1/2\left(\frac{d\phi}{d\xi}\right)^2$; its total energy, $H = 1/2\left(\frac{d\phi}{d\xi}\right)^2 + \left(E\phi^2 + 1/2\alpha\phi^4\right)$, is a constant of motion. Two typical plots of the potential function in the cases of $E > 0$ and $E < 0$ are illustrated in Fig. 3.1(a). As shown, both plots represent potential wells.

In the case of $E > 0$, the total energy H of the trapped object is larger than zero, i.e., $H > 0$; the trapped object is bounced back and forth in the potential well to have an oscillatory trajectory $\phi_{p1}(\xi)$, which is a symmetric alternate function illustrated in Fig. 3.1(b). In the case of $E < 0$, the trapped object has $H > 0$ or $E\phi_m^2 + 1/2\alpha\phi_m^4 < H < 0$. If $H > 0$, the bounce motion of the object has an oscillatory trajectory $\phi_{p2}(\xi)$, which is a symmetric alternate function illustrated in Fig. 3.1(b). If $E\phi_m^2 + 1/2\alpha\phi_m^4 < H < 0$, the oscillatory trajectory $\phi_{p3}(\xi)$ of the object is a non-alternate periodic function as shown in Fig. 3.1(b). Moreover, in the case of $E < 0$, there exists a non-oscillatory trajectory with $H = 0$. Let the object start at $\phi = \phi_1(= 0)$ and examine the motion of the object in the region between ϕ_1 and ϕ_2 in Fig. 3.1(a). Initially, it moves very slowly to the right. As it drops into the potential well, it moves quickly toward the potential minimum at φ_m. After passing the potential minimum, the object starts to climb up to the turning point at $\phi = \phi_2$, where the kinetic energy T of the object reduces to zero; it is then bounced back into

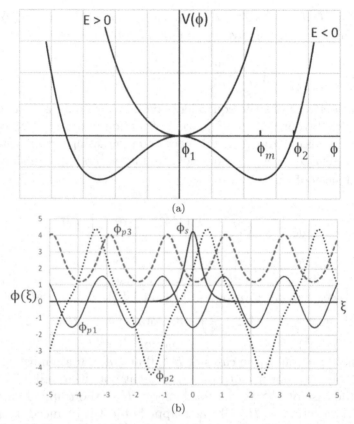

Fig. 3.1.　(a) Two potential distributions showing three kinds of potential wells and (b) three periodic solutions ϕ_{p1}, ϕ_{p2}, and ϕ_{p3}, corresponding to three periodic trajectories bouncing in the respective potential wells shown in (a), and a solitary solution ϕ_s, corresponding to an aperiodic trajectory which only bounces once in the $E < 0$ potential well.

the potential well. It quickly passes the potential minimum and then climbs up toward the starting point $\phi_1 = 0$. It takes a long time for the object to reach ϕ_1, where the object stays. This non-oscillatory trajectory $\phi_s(\xi)$ is also shown in Fig. 3.1(b) and represents a solitary solution of (1.24).

3.1.2.1　*Periodic solutions*

The potential distribution functions illustrated in Fig. 3.1(a) indicate that the solutions of (3.2) in the case of $H \neq 0$ are periodic as illustrated in Fig. 3.1(b). The analytical solutions of (3.2) are found in special cases as exemplified in the following:

1. For $E > 0$, i.e., $H > 0$. Set $\eta_1 = \left[2E/\left(1 - 2k_1^2\right)\right]^{1/2}\xi$, $\phi(\xi) = \phi_{10}y_1(\eta_1)$, and $k_1^2 = \left(1 - 2k_1^2\right)\frac{\alpha\phi_{10}^2}{2E} = \frac{1}{2}\left(1 - \frac{1}{\sqrt{1+2\alpha H/E^2}}\right) < \frac{1}{2}$, where $\phi_{10}^2 = \frac{E}{\alpha}\left(\sqrt{1 + 2\alpha H/E^2} - 1\right)$, (3.2) is normalized as

$$y_1'' + \left(1 - 2k_1^2\right)y_1 + 2k_1^2y_1^3 = 0 \qquad (3.4a)$$

where $y_1'' = \frac{d^2}{d\eta_1^2}y_1$. The solution of (3.4a) is a Jacobi elliptic (cosine amplitude) function $\mathrm{cn}(\eta_1, k_1)$; thus,

$$\phi_{p1}(\xi) = \phi_{10}\mathrm{cn}(\eta_1, k_1)$$

which is a symmetric alternate function consistent with the ϕ_{p1} plot shown in Fig. 3.1(b).

2. For $E < 0$, $H > 0$. Again, set $\eta_2 = \left[2E/(1 - 2k_2^2)\right]^{1/2}\xi$, $\phi(\xi) = \phi_{20}y_2(\eta_2)$, and $k_2^2 = \left(1 - 2k_2^2\right)\frac{\alpha\phi_{20}^2}{2E} = \frac{1}{2}\left(1 + \frac{1}{\sqrt{1+2\alpha H/E^2}}\right) > \frac{1}{2}$, where $\phi_{20}^2 = -\frac{E}{\alpha}\left(\sqrt{1 + 2\alpha H/E^2} + 1\right)$, (3.2) is normalized as

$$y_2'' + \left(1 - 2k_2^2\right)y_2 + 2k_2^2y_2^3 = 0 \qquad (3.4b)$$

where $y_2'' = \frac{d^2}{d\eta_2^2}y_2$. Equation (3.4b) has the same form as (3.4a), its solution is also a Jacobi elliptic (cosine amplitude) function $\mathrm{cn}(\eta_2, k_2)$. Thus,

$$\phi_{p2}(\xi) = \phi_{20}\mathrm{cn}(\eta_2, k_2)$$

which is represented by a symmetric alternate function ϕ_{p2} shown in Fig. 3.1(b). It is noted that ϕ_{p1} and ϕ_{p2} have different function forms due to the shape difference of the two wells; ϕ_{p2} has a larger amplitude and period due to bouncing in a wider well.

3. For $E < 0$ and $H < 0$. Set $\eta_3 = [-2E/(2 - k_3^2)]^{1/2}\xi$, $\phi(\xi) = \phi_{30}y_3(\eta_3)$, and $-(2 - k_3^2)\frac{\alpha\phi_{30}^2}{E} = 2$, i.e., $k_3^2 = 2(1 + \frac{1}{\sqrt{1+2\alpha H/E^2}})^{-1}$, where $\phi_{30}^2 = -\frac{E}{\alpha}(\sqrt{1 + 2\alpha H/E^2} + 1)$, (3.2) is normalized as

$$y_3'' - (2 - k_3^2)\, y_3 + 2y_3^3 = 0 \tag{3.4c}$$

where $y_3'' = \frac{d^2}{d\eta_3^2}y_3$. The solution of (3.4c) is a Jacobi elliptic (delta amplitude) function $dn(\eta_3, k_3)$; thus,

$$\phi_{p3}(\xi) = \phi_{30}\, dn(\eta_3, k_3)$$

which is a non-alternate periodic function consistent with the ϕ_{p3} plot shown in Fig. 3.1(b).

3.1.2.2 *Solitary solution*

A localized solution of (3.2) requires $\phi = 0 = d\phi/d\xi$ as $|\xi| \to \infty$, thus, $H = 0$ in the case of $E < 0$ is considered. Set $x = \sqrt{2|E|}\xi$ and $\phi(\xi) = \phi_{s0}Y_s(x)$, where $\phi_{s0} = \sqrt{\frac{2|E|}{\alpha}}$, (3.2) is normalized as

$$Y_s'' - Y_s + 2Y_s^3 = 0 \tag{3.5a}$$

where $Y_s'' = \frac{d^2}{dx^2}Y_s$. Compare (3.5a) with (3.4b) and (3.4c), the solution of (3.5a) is $cn(x, 1) = dn(x, 1) = \text{sech}\, x$. Thus,

$$\phi_s(\xi) = \sqrt{\frac{2|E|}{\alpha}}\text{sech}\sqrt{2|E|}\xi \tag{3.5b}$$

This is a solitary solution that $\phi_s = 0 = d\phi_s/d\xi$ as $|\xi| \to \infty$. Its width ($\propto \frac{1}{\sqrt{2|E|}}$) is inversely proportional to its amplitude. A representation of this solution is also plotted in Fig. 3.1(b).

Exercise 3.6: Show that $\text{sech}x$ is a solution of (3.5a).

The cubic nonlinearity of the medium mitigates wave dispersion in the propagation; when the nonlinear effect ($\propto$ square of the amplitude) and the dispersion effect (inversely proportional to the square of the width) reach a balance, a shape-preserved solitary wave is formed and trapped in the self-induced density well ($\propto -\phi_s^2$).

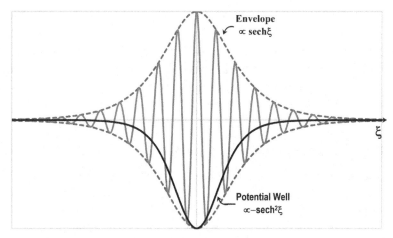

Fig. 3.2. Envelope soliton solution of the nonlinear Schrödinger equation (1.24) and the self-induced potential well trapping the soliton.

With the aid of (3.5b), the field function of a solitary wave packet is obtained as

$$E_s(z, t) = \sqrt{\frac{2\,|E|}{\alpha_0}}\,\text{sech}\left[\sqrt{2\,|E|}k_0\,(z - v_g t)\right]$$

$$\times\,\cos[k_0 z - (\omega_0 + \Delta\omega)\,t] \qquad (3.6)$$

where $\alpha_0 = \frac{\alpha}{|\psi_0|^2}$ and $\Delta\omega = Ek_0 v_g < 0$. This envelope soliton (3.6) and the self-induced potential well ($\propto -\phi_s^2$) to trap this solitary wave are plotted in Fig. 3.2.

With the aid of the relations

$$\int_{-\infty}^{\infty} \text{sech}\,z dz = \pi \text{ and } \int_{-\infty}^{\infty} \text{sech}^2 z dz = 2$$

it is shown that the area $\int_{-\infty}^{\infty} \phi_s(\xi)d\xi = \frac{\pi}{\sqrt{\alpha}}$ of a soliton is independent of the amplitude; moreover, its energy $\int_{-\infty}^{\infty} \phi_s^2(\xi)d\xi = 2\frac{\sqrt{2|E|}}{\alpha}$ is linearly proportional to the amplitude. This is because the width of the soliton is inversely proportional to the amplitude. Both quantities are inversely proportional to the square root of the nonlinear coefficient α (i.e., $\propto \frac{1}{\sqrt{\alpha}}$).

The solution (3.6) indicates that plasma can support solitons, which are shape-preserved localized wave packets. However, it is realized that soliton is not a necessity of nonlinearity; for instance, the nonlinear plasma waves, in general, are periodic; soliton(s) may appear when the source wave function has a localized form, for instance, a Gaussian pulse shape.

3.1.3   *Collapse of nonlinear waves*

Intense electrostatic waves can be excited in plasma, for example, by the parametric instabilities and by the electron beam–plasma instability. Depending on the operation mode of the driver, the excited plasma waves can be periodic waves or localized wave packets. The nonlinearity of the plasma gives rise to mode coupling. It broadens the spatial spectrum of the plasma waves, which diffuse tail electrons in the velocity distribution via quasi-linear and further resonance broadening mechanisms. On the other hand, plasma wave packets may be generated by the localized drivers or in the localized resonance region, e.g., upper hybrid waves in the upper hybrid resonance region discussed in Section 2.5 of Chapter 2; their propagation is governed by a nonlinear Schrödinger equation (2.7). This equation is analyzed in the one-dimensional case for a stationary solution and for eigenstates. However, (2.7) can be extended to the multidimensional cases. In multidimensional systems, the solution of the equation may not be stable; one way to explore the stability conditions is via the "Virial Theorem".

Extend the variance $V(\tau)$ of (3.1g) for a one-dimensional system to an n-dimensional system, it becomes

$$V(\tau) = \int |\mathbf{r}|^2 |\varphi(\mathbf{r}, \tau)|^2 d\mathbf{r} \qquad (3.7)$$

The variance identity becomes

$$\frac{d^2}{d\tau^2} V(\tau) = 4H + (2 - n)\alpha \int |\varphi(\mathbf{r}, \tau)|^4 \, d\mathbf{r} \qquad (3.8)$$

Equation (3.8) is integrated as

$$V(\tau) = V(0) + \tau \frac{d}{d\tau} V(0) + 2H\tau^2$$

$$+ (2 - n)\alpha \int_0^\tau \left[(\tau - s) \int |\varphi(\mathbf{r}, s)|^4 \, d\mathbf{r} \right] ds \qquad (3.9)$$

where $V(\tau) \geq 0$ for all τ;

$$\frac{d}{d\tau}V(0) = i \int |\varphi_0|^2 \, \mathbf{r} \cdot \nabla \left(\ln \frac{\varphi_0^*}{\varphi_0} \right) d\mathbf{r} = 2 \int \mathbf{r} \cdot J_{10} \, d\mathbf{r}$$

and J_{10} is the initial value of the momentum density J_1 defined in Section 3.1.1.1. If the RHS can become negative at finite time τ, it suggests that the solution collapses. When $n \geq 2$, the last term on the RHS of (3.9) is zero for $n = 2$ and negative for $n = 3$. Therefore, if $H < 0$, collapse can occur.

Exercise 3.7: Show that (3.9) satisfies (3.8).

In an isotropic plasma, $\nabla^2 = \frac{1}{r^2}\frac{d}{dr}r^2\frac{d}{dr}$, where r is the radial coordinate in the spherical coordinates, (2.7) is re-expressed as

$$-1/2\frac{1}{r^2}\frac{d}{dr}r^2\frac{d}{dr}E - \alpha_1|E|^2 E = i\frac{\partial}{\partial t}E \tag{3.10}$$

A solution with a localized form, i.e., trapped in a self-induced caviton, is considered. In the following, (3.10) is solved for approximate solution in the two regions: $r < \frac{1}{2}$ and $r > 1$.

1. $r < \frac{1}{2}$
Set $E(r,t) = \frac{E_0}{r^2+1}e^{-i\lambda t}$, where λ is a real eigenvalue of a state; each term of (3.10) becomes

$$\tfrac{1}{2}\frac{1}{r^2}\frac{d}{dr}r^2\frac{d}{dr}E = E_0 e^{-i\lambda t}\left[\frac{1}{(r^2+1)^2} - \frac{4}{(r^2+1)^3}\right]$$

$$\alpha_1|E|^2 E = \frac{\alpha_1 E_0^3}{(r^2+1)^3}e^{-i\lambda t}$$

$$i\frac{\partial}{\partial t}E = \frac{\lambda E_0}{r^2+1}e^{-i\lambda t}$$

It shows that, with $E_0 = 2/\sqrt{\alpha_1}$ and $\lambda = -1$, $E(r,t) = \frac{E_0}{r^2+1}e^{it}$ is a good, approximated solution of (3.10) for $r < 1/2$.

2. $r > 1$

Set $E(r,t) = \frac{E_1 e^{-\beta r}}{r} e^{it}$; each term of (3.10) becomes

$$1/2 \frac{1}{r^2} \frac{d}{dr} r^2 \frac{d}{dr} E = \frac{\beta^2 E_1}{2r} e^{-\beta r} e^{it}$$

$$\alpha_1 |E|^2 E = \frac{\alpha_1 E_1^3}{r^3} e^{-3\beta r} e^{it}$$

$$i \frac{\partial}{\partial t} E = -\frac{E_1}{r} e^{-\beta r} e^{it}$$

With $\beta = \sqrt{2}$, the cubic nonlinear term, $\alpha_1 |E|^2 E$, can be neglected; thus, $E(r,t) = \frac{E_1 e^{-\sqrt{2} r}}{r} e^{it}$ is a good, approximated solution of (3.10) for $r > 1$.

Equal two solutions at $r = 1$, it leads to $E_1 = E_0 e^{\sqrt{2}}/2$.

Substitute $E = A(r,t)/r$ into (3.10), it yields

$$-1/2 \frac{d^2}{dr^2} A - \frac{\alpha_1 |A|^2 A}{r^2} = i \frac{\partial}{\partial t} A \tag{3.11}$$

It shows that, in the isotropic case, (3.10) is reduced to a one-dimensional nonlinear Schrödinger equation; however, the nonlinear potential $V(r,t) = -\alpha_1 |A|^2/r^2$ is proportional to the wave intensity ($|A|^2$) as well as to the dimensional effect ($\propto 1/r^{n-1}$); in other words, the dimensionality of the system is yet preserved.

As the excited waves coalesce into localized wave packets (condensation and nucleation phase), the induced ponderomotive forces push out local plasma to generate density cavities (with $H < 0$). Due to $1/r^2$ dependent, the nonlinear wave functions will continue to steepen, and density cavities become deeper and more local (localization phase); the process does not reach a steady state; instead, the steepened nonlinear waves collapse into short wavelength propagating waves (i.e., the steepened nonlinear waves become localized sources to generate new waves). These waves suffer significant Landau damping (dissipation and burnout phase) by the bulk elections in the velocity distribution. On the other hand, the mode coupling process broadens the spatial spectrum of a linear Langmuir wave into the long wavelength regime. Those waves, having large phase velocities, facilitate quasi-linear and resonance-broadening diffusion of the tail electrons in the velocity distribution. After burnout, the remaining waves relax into linear waves, which are then amplified by the source to repeat the cycle. Numerical simulations have illustrated such a nonlinear wave collapse process, the results show that the velocity and density distributions of plasma are modified significantly.

3.2 Korteweg–de Vries (KdV) Equation

It models a variety of nonlinear phenomena, including ion acoustic waves in plasmas and shallow water waves. In (1.1), the first term characterizes the time evolution of the wave propagating in one direction, the second term disperses the wave, and the third term steepens the wave.

3.2.1 *Conservation laws*

The KdV equation (1.1) also possesses many conservation laws, in the form of the continuity equation $\frac{\partial}{\partial \tau} N + \frac{\partial}{\partial \eta} J = 0$, where N is the density of a conserved quantity and J is the flux of this quantity. The densities N of the first two conservation laws are in the form of ϕ^n, for $n = 1$ and 2; in the higher order cases, N has a more complicated form, e.g., $N = \phi^3 - \frac{1}{2}\phi_\eta^2$ in the third conservation law. In the following, the first three conservation laws are illustrated. Equation (1.1) is rearranged to obtain the first conservation law

$$\frac{\partial}{\partial \tau}\phi + \frac{\partial}{\partial \eta}(3\phi^2 + \phi_{\eta\eta}) = 0 \tag{3.12a}$$

Next, multiplying (1.1) by ϕ, the second conservation law is obtained as

$$\frac{\partial}{\partial \tau}\frac{\phi^2}{2} + \frac{\partial}{\partial \eta}\left[2\phi^3 + \phi\phi_{\eta\eta} - \frac{1}{2}\phi_\eta^2\right] = 0 \tag{3.12b}$$

The third conservation law is given as

$$\frac{\partial}{\partial \tau}\left(\phi^3 - \frac{1}{2}\phi_\eta^2\right) + \frac{\partial}{\partial \eta}\left[\frac{9}{2}\phi^4 + 3\phi^2\phi_{\eta\eta} - 6\phi\phi_\eta^2 + \frac{1}{2}\phi_{\eta\eta}^2 - \phi_\eta\phi_{\eta\eta\eta}\right]$$
$$= 0 \tag{3.12c}$$

Higher-order conservation laws can be found in literature. It is conjectured that the KdV equation has an infinite number of conservation laws.

3.2.2   *Potential and modified Korteweg–de Vries (pKdV and mKdV) equations*

There are several variants of the KdV equation, which are derived via different function transformations:

1. Set $\phi(\eta, \tau) = \frac{\partial}{\partial \eta} U_p(\eta, \tau)$ in (1.1), it leads to the "pKdV" equation

$$\frac{\partial}{\partial \tau} U_p + \frac{\partial^3}{\partial \eta^3} U_p + 3 \left(\frac{\partial}{\partial \eta} U_p \right)^2 = 0 \qquad (3.13a)$$

It is the same as (1.2a), which has a solution (1.3g) re-expressed as

$$U_p = 2\kappa \tanh[\kappa(\eta - 4\kappa^2 \tau) + \theta_0] \qquad (3.13b)$$

2. Apply Miura transformation

$$\phi = - \left(U^2 + \frac{\partial}{\partial \eta} U \right) \qquad (3.14)$$

to the KdV equation (1.1), i.e., substitute (3.14) into (1.1), it shows that $U(\eta, \tau)$ satisfies an mKdV equation

$$\frac{\partial}{\partial \tau} U + \frac{\partial^3}{\partial \eta^3} U - 6U^2 \frac{\partial}{\partial \eta} U = 0 \qquad (3.15a)$$

Set $U = i\psi$, (3.15a) becomes

$$\frac{\partial}{\partial \tau}\psi + \lambda^2 \frac{\partial}{\partial \eta}\psi + \frac{\partial}{\partial \eta} \left[\frac{\partial^2}{\partial \eta^2}\psi + (-\lambda^2 + 2\psi^2)\psi \right] = 0 \quad (3.15b)$$

For a propagating mode, $\psi(\eta, \tau) = \psi(\xi)$, where $\xi = \eta - \lambda^2 \tau + \theta$; θ is a constant parameter; then,

$$\frac{\partial}{\partial \tau}\psi + \lambda^2 \frac{\partial}{\partial \eta}\psi = 0$$

and

$$\frac{\partial^2}{\partial \eta^2}\psi + (-\lambda^2 + 2\psi^2)\psi = 0 \qquad (3.15c)$$

It is noted that (3.15c) is the same as (3.2) with $\alpha = 1$ and $E = -\lambda^2/2$. Thus, a propagating solitary solution of (3.15a) is given as

$$U = \pm i\lambda \operatorname{sech} \lambda(\eta - \lambda^2 \tau + \theta) \qquad (3.15d)$$

Both (3.13a) and (3.15a) also possess many conservation laws.

Exercise 3.8: Find the first conservation laws of the pKdV (3.13a) and mKdV (3.15a) equations.

Ans:

$$\frac{\partial}{\partial \tau} U_{p\eta} + \frac{\partial}{\partial \eta}(U_{p\eta\eta\eta} + 3U_{p\eta}^2) = 0$$

$$\frac{\partial}{\partial \tau} U^2 + \frac{\partial}{\partial \eta}(2UU_{\eta\eta} - U_\eta^2 - 3U^2) = 0$$

3.2.3 *Propagating modes*

Consider traveling wave solution of the form $\phi(\eta, \tau) = f(\xi)$, where $\xi = \eta - A_s\tau$, (1.1) becomes

$$\frac{d^3}{d\xi^3} f - A_s \frac{d}{d\xi} f + 3\frac{d}{d\xi} f^2 = 0 \qquad (3.16a)$$

This equation is integrated as

$$\frac{d^2}{d\xi^2} f - A_s f + 3f^2 = C_0 \qquad (3.16b)$$

where the integration constant $C_0 = [f''(0) - A_s f(0) + 3f^2(0)]$. It is noted that C_0 can be canceled by shifting f by a constant, i.e., $f = f_1 + C_1$ and $C_1 = \frac{A_s}{6}(1 - \sqrt{1 + 12C_0/A_s^2})$; on the other hand, C_1 has to be zero physically; hence, $C_0 = 0$.

Again, consider (f, ξ) as the coordinate and time of a one-dimensional space, like (3.2), (3.16b) also represents an equation of motion of a unit mass object moving in this one-dimensional space and interacting with a restoring force of $A_s f - 3f^2$.

Multiply $\frac{d}{d\xi}f$ with (3.16b), yields an invariant equation

$$\frac{d}{d\xi}\left[\frac{1}{2}\left(\frac{d}{d\xi}f\right)^2 + \left(-\frac{A_s}{2}f^2 + f^3\right)\right] = 0 \qquad (3.16c)$$

It indicates that the quantity in the parentheses on the LHS of (3.16c) is invariant to ξ; that is,

$$\tfrac{1}{2}f'^2 + (-\tfrac{1}{2}A_s f^2 + f^3) = H \qquad (3.16d)$$

where $f' = df/d\xi$ and the integration constant $H = \tfrac{1}{2}[f'^2(0) - A_s f^2(0) + 2f^3(0)]$.

Thus, this unit mass object is moving in a potential field, having potential energy $VE = (-\tfrac{1}{2}A_s f^2 + f^3)$, kinetic energy $KE = \tfrac{1}{2}f'^2$, and a constant total energy $TE = VE + KE = H$. A typical plot of the potential field $V(f) = VE$, for $A_s = 2$, is illustrated in Fig. 3.3(a). It exhibits a potential well, which traps objects to force oscillations.

The oscillatory trajectories convert to periodic solutions of (3.16a). As an example, a periodic solution, for $H = -3/64$ and the initial conditions, $f(0) = 2/3$ and $f'(0) = 0$, is illustrated in Fig. 3.3(b). Moreover, there exists a non-oscillatory trajectory in the case of $H = 0$. It will be shown as a solitary solution of (3.16a).

3.2.3.1   *Periodic solution*

Although (3.16a) cannot be converted into a Jacobi elliptic equation, its periodic solution, under a special condition, can be obtained indirectly via the mKdV equation (3.15a). For the traveling wave solution of the form $U(\eta, \tau) = g(\xi)$, where $\xi = \eta - A_s \tau + \theta$, (3.15a) becomes

$$\frac{d^3}{d\xi^3}g - A_s\frac{d}{d\xi}g - 2\frac{d}{d\xi}g^3 = 0 \qquad (3.17a)$$

This equation is integrated as

$$\frac{d^2}{d\xi^2}g - A_s g - 2g^3 = C \qquad (3.17b)$$

Consider a special case that the integration constant $C = 0$ and set $g(\xi) = \pm iy(\xi)$ and $A_s = (2 - k^2)$, then (3.17b) is converted into a

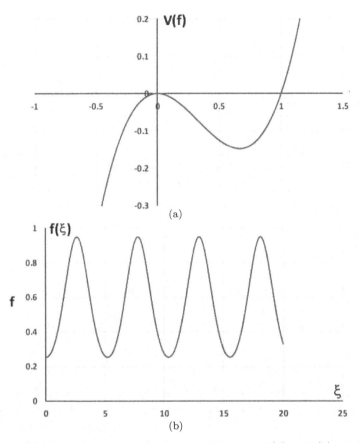

Fig. 3.3. (a) A typical plot of the potential function $V(f)$ and (b) a periodic solution of (3.16a).

Jacobi elliptic equation

$$y'' - (2 - k^2)y + 2y^3 = 0 \qquad (3.17c)$$

Its solution is the Jacobi elliptic (sine amplitude) function $\mathrm{cn}(\xi, k)$ Hence, the periodic solution of (3.16a) in the case of $C = 0$ is obtained as

$$f(\xi; k) = \mathrm{cn}^2(\xi, k) \pm i\,\mathrm{sn}(\xi, k)\,\mathrm{dn}(\xi, k) \qquad (3.16e)$$

where the constant parameter k is determined by the initial condition.

For $k \neq 1$, $f(\xi; k)$ is a periodic function, for instance,

$$f(\xi; 0) = \cos^2(\eta - 2\tau + \theta) \pm i\sin(\eta - 2\tau + \theta) \qquad (3.16f)$$

It is a periodic solution of the KdV equation.

In the case of $k = 1$, (3.16e) becomes

$$f(\xi; 1)$$
$$= \text{sech}^2(\eta - \tau + \theta) \pm i\,\text{sech}(\eta - \tau + \theta)\tanh\lambda(\eta - \lambda^2\tau + \theta)$$
$$= \text{sech}(\eta - \tau + \theta)\exp[\pm i\tan^{-1}\sinh(\eta - \tau + \theta)] \qquad (3.16g)$$

This is a solitary solution.

3.2.3.2 *Soliton trapped in self-induced potential well*

Equation (3.16d), with $H = 0$, leads to $f' = -f\sqrt{A_s - 2f}$, which is then integrated to obtain a solitary solution of the form

$$f(\xi) = \phi(\eta, \tau) = \phi(\eta - A_s\tau) = \tfrac{1}{2}A_s\,\text{sech}^2[\tfrac{1}{2}\sqrt{A_s}(\eta - A_s\tau)] \qquad (3.18a)$$

which is the same as (1.4a) with $A_s = 2\beta$.

It is shown that the velocity (in the moving frame) of an ion acoustic soliton is proportional to its amplitude A_s and the width is inversely proportional to the square root of the amplitude. This is realized because, in (3.12a), the nonlinear effect is proportional to the amplitude and the dispersion effect is inversely proportional to the square of the width.

At the identified amplitude–width relationship, these two effects reach balance to form a shape-preserved soliton, which is trapped in the self-induced density well ($\propto -\phi_s^2$). The area $\int_{-\infty}^{\infty} \phi(\eta, \tau)d\eta = 2A_s^{1/2}$ and energy $\int_{-\infty}^{\infty} \phi^2(\eta, \tau)d\eta = \tfrac{2}{3}A_s^{3/2}$ of an ion acoustic soliton are amplitude-dependent. These relations impose conditions on the source (initial) pulse which is likely to evolve nonlinearly to become a soliton as shown in Fig. 3.4, in which the self-induced ponderomotive force to balance the wave dispersion is also plotted.

An ion acoustic soliton can be seen as a traveling plasma density bump. Substitute (3.18a) into the ion continuity equation of (2.1) $\frac{\partial}{\partial t}\frac{n_{si}}{n_0} + \frac{\partial}{\partial z}V_{si} = 0$, where $V_{si} = \phi/\alpha_2$ (refer to Section 2.3 of

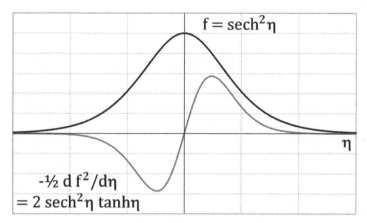

Fig. 3.4. Ion acoustic soliton and the self-induced ponderomotive force to balance the dispersion.

Chapter 2), and $\tau = \omega_{pi}t$ and $\eta = \beta(z - C_s t)$, this traveling density bump has the form

$$\frac{n_{si}}{n_0} = \frac{12A_{s1}}{1 + 2A_{s1}}\text{sech}^2\left[\left(\frac{C_s}{v_s}\right)\sqrt{A_{s1}}k_{De}(z - C_{s1}t)\right] \quad (3.18b)$$

where $A_{s1} = (v_s/2C_s)^{4/3}A_s$ and $C_{s1} = (1 + 2A_{s1})C_s$.

Exercise 3.9: Verify that (3.16f) and (3.16g) are solutions of the KdV equation (3.16a).

3.2.4 *Solitary solution with Miura transform*

Since the mKdV equation (3.15a) is derived from the KdV equation (1.1) via Miura transformation (3.14), its solution (3.15d) can be transformed back to obtain a solitary solution for the KdV equation

$$\phi_1 = -\left(U^2 + \frac{\partial}{\partial\eta}U\right)$$

$$= \lambda^2[\text{sech}^2\lambda(\eta - \lambda^2\tau + \theta)$$

$$\pm i\,\text{sech}\,\lambda(\eta - \lambda^2\tau + \theta)\tanh\lambda(\eta - \lambda^2\tau + \theta)]$$

$$= \lambda^2\text{sech}\,\lambda(\eta - \lambda^2\tau + \theta)$$

$$\times \exp[\pm i\tan^{-1}\sinh\lambda(\eta - \lambda^2\tau + \theta)] \quad (3.19a)$$

Then,

$$|\phi_1| = \lambda^2 \operatorname{sech} \lambda(\eta - \lambda^2 \tau + \theta) \tag{3.19b}$$

It is noted that (3.19a) is a general form of (3.16g). It is realized that both solutions are derived via Miura transformation and mKdV equation. Set $1/2\sqrt{A_s} = \lambda$, it shows that the soliton (3.19a) is wider but lower than the soliton (3.18a).

3.3   Burgers Equation

Burgers equation describes nonlinear wave motion that has undergone linear diffusion and is the simplest model for analyzing the combined effect of nonlinear advection and diffusion. On the LHS of (1.8a), the first term characterizes the time evolution of the wave propagating in one direction and the second term induces nonlinear convection; the RHS term induces linear diffusion.

The nonlinear nature of Burgers equation has been exploited as a useful prototype differential equation for modeling many phenomena, such as shock flows, wave propagation in combustion chambers, vehicular traffic movement, acoustic transmission. It is applied to study flow phenomenon regarding the balancing effects of viscous and inertial or convective forces. When inertia or convective forces are dominant, its solution resembles that of the kinematic wave equation which displays a propagating wave front and boundary layers. In contrast, when viscous forces are dominant, the propagating wave front is smeared and diffused due to viscous action.

3.3.1   *Analytical solution via the Cole–Hopf transformation*

The initial value problem for the viscid Burgers equation (1.8a) is solved analytically by introducing the Cole–Hopf transformation

$$\phi = -2b\frac{\varphi_x}{\varphi} = -2b\frac{\partial}{\partial x}\ln\varphi \tag{3.20}$$

Operating on (3.20) for the terms in (1.8a) yields

$$\phi_t = 2b\frac{\varphi_t\varphi_x - \varphi\varphi_{tx}}{\varphi^2}, \quad \phi\phi_x = 4b^2\frac{\varphi_x(\varphi\varphi_{xx} - \varphi_x^2)}{\varphi^3}, \quad \text{and}$$

$$b\phi_{xx} = -2b^2\frac{\varphi^2\varphi_{xxx} - 3\varphi\varphi_x\varphi_{xx} + 2\varphi_x^3}{\varphi^3}$$

With the aid of these relations, (1.8a) becomes

$$2b\frac{-\varphi\varphi_{tx} + \varphi_x(\varphi_t - b\varphi_{xx}) + b\varphi\varphi_{xxx}}{\varphi^2} = 0 \qquad (3.21a)$$

It leads to

$$\varphi_x(\varphi_t - b\varphi_{xx}) = \varphi(\varphi_{tx} - b\varphi_{xxx}) = \varphi(\varphi_t - b\varphi_{xx})_x \qquad (3.21b)$$

Equation (3.21b) shows that φ is governed by a linear diffusion equation

$$\varphi_t - b\varphi_{xx} = 0 \qquad (3.22)$$

where b is the diffusion coefficient. This equation is the same as (1.8e). The general solution of the diffusion equation is well known, however, in the current case, its initial condition $\varphi(x,0) = \varphi_0(x)$ is imposed by the initial condition $\phi(x,0) = \phi_0(x)$ of the Burgers equation (1.8a) and is determined via the same transformation (3.20) that is integrated as

$$\varphi(x,t) = \exp\left(-\frac{1}{2b}\int_{-\infty}^{x}\phi(\xi,t)\,d\xi\right) \qquad (3.23a)$$

Then, the initial condition of the diffusion equation (3.22) is given as

$$\varphi_0(x) = \exp\left(-\frac{1}{2b}\int_{-\infty}^{x}\phi_0(\xi)\,d\xi\right) \qquad (3.23b)$$

The diffusion equation (3.22) is solved by taking the Fourier transform with respect to x and then integrating on t; it yields

$$\hat{\psi}(k,t) = \hat{\psi}_0(k)e^{-bk^2t} \qquad (3.24)$$

where the Fourier transform is defined as

$$\hat{\psi}(k,t) = \int_{-\infty}^{\infty}\varphi(x,t)e^{-ikx}\,dx$$

The solution $\varphi(x,t)$ of (3.22) is the inverse Fourier transform of (3.24). Since the RHS of (3.24) is a product of two transform functions, the inverse Fourier transform of (3.24) is the convolution of those two functions in the x space (the inverse Fourier transform of $\hat{\psi}_0(k)$ and e^{-bk^2t}). The inverse Fourier transform of e^{-bk^2t} is evaluated as

$$\int_{-\infty}^{\infty} e^{-bk^2t} e^{ikx} \frac{dk}{2\pi} = \frac{1}{2\sqrt{\pi bt}} e^{-\frac{x^2}{4bt}} = g(x,t)$$

Then,

$$\varphi(x,t) = \int_{-\infty}^{\infty} \hat{\psi}(k,t) e^{ikx} \frac{dk}{2\pi} = \varphi_0(x) \circ g(x,t)$$

$$= \frac{1}{2\sqrt{\pi bt}} \int_{-\infty}^{\infty} \varphi_0(\xi) e^{-\frac{(x-\xi)^2}{4bt}} d\xi \tag{3.25}$$

where "$\circ$" denotes the convolution product. Substitute (3.25) into (3.20), the analytical solution of (1.8a) is obtained as

$$\phi(x,t) = \frac{\int_{-\infty}^{\infty} \frac{(x-\xi)}{t} \varphi_0(\xi) e^{-\frac{(x-\xi)^2}{4bt}} d\xi}{\int_{-\infty}^{\infty} \varphi_0(\xi) e^{-\frac{(x-\xi)^2}{4bt}} d\xi}$$

$$= -2b \frac{\partial}{\partial x} \ln \varphi(x,t) \tag{3.26a}$$

Exercises 1.8 and 1.9 of Chapter 1 show that $\varphi_1(x,t) = \frac{1}{\sqrt{1+4bt}} \exp\left(-\frac{x^2}{1+4bt}\right)$ and $\varphi_2(x,t) = 1 + \frac{1}{\sqrt{t}} \exp\left(-\frac{x^2}{4bt}\right)$ are solutions of the diffusion equation (3.22); substitute those into (3.26a), two analytical solutions of the Burgers equation (1.8a) are derived, as examples, as

$$\phi_1(x,t) = \frac{4bx}{1+4bt} \tag{3.26b}$$

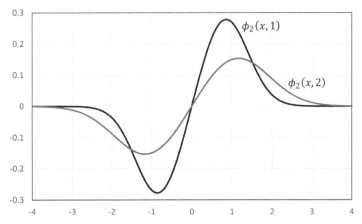

Fig. 3.5. Plots of N-wave solution of the Burgers equation (1.8a) for $b = 0.25$ and $t = 1$ and 2.

and

$$\phi_2(x,t) = \frac{\frac{x}{t}}{1 + \sqrt{t}\exp\left(\frac{x^2}{4bt}\right)} \qquad (3.26c)$$

Solution (3.26c) is commonly called "N-wave" solution because it has an N-shaped graph at any time (for $t > 0$), as shown in Fig. 3.5, with two representing plots for $b = 0.25$ and $t = 1$ and 2.

Exercise 3.10: Find the solutions of the Burgers equation (1.8a) for the initial condition $\varphi_0(x) = \cos\frac{\pi x}{L}$ of the diffusion equation (3.22).

Ans: $\phi(x,t) = \frac{2b\pi}{L}\tan\frac{\pi x}{L}$.

3.3.2 *Propagating modes*

Consider traveling wave solution of the form $\phi_1(\eta, \tau) = F_1(\xi)$, where $\xi = \eta - A_b\tau$, (2.10) (which is the same as (1.8a)) becomes

$$b\frac{d^2}{d\xi^2}F_1 + \frac{d}{d\xi}\left(A_bF_1 - \frac{1}{2}F_1^2\right) = 0 \qquad (3.27a)$$

It is integrated as

$$b\frac{d}{d\xi}F_1 + \left(A_bF_1 - \frac{1}{2}F_1^2\right) = 0 \qquad (3.27b)$$

where the integration constant is set to zero for the same reason that it can be canceled by a constant shift of F_1. Equation (3.27b) is integrated to obtain

$$F_1(\xi) = A_b \left(1 - \tanh \frac{A_b}{2b} \xi \right) \tag{3.28}$$

It shows that F_1 starts at a level of $2A_b$, i.e., $F_1(-\infty) = 2A_b$, and decreases continuously to zero as it moves to ∞, i.e., $F_1(\infty) = 0$. The transition width is proportional to $2b/A_b$, i.e., proportional to the damping factor b and inversely proportional to the amplitude A_b. As $b \to 0$, it becomes a shock wave, like that shown in Fig. 1.1(d). The shock front has a step transition at $\xi = 0$.

Problems

P3.1. In linear system, the Schrödinger eigenvalue equation (3.2) becomes

$$\left(-1/2 \frac{d^2}{d\xi^2} + V(\xi) \right) \phi = E\phi \tag{P3.1}$$

where $V(\xi)$ is the potential function of the system. Assume that $V(\xi) = -k^2 \text{sech}^2(k\xi)$, find the solitary eigenfunction $\phi(\xi)$ and the corresponding eigenvalue E.

P3.2. In Problem P3.1, change the potential function to $V(\xi) = -\exp(-2|\xi|)$; show that with the transform $z = \sqrt{2}e^{\xi}$ for $\xi < 0$ and $z = \sqrt{2}e^{-\xi}$ for $\xi > 0$, the function $y(z) = \phi(\xi)$ satisfies a Bessel equation, i.e., (P3.1) is transformed into the Bessel differential equation

$$\frac{\partial^2}{\partial z^2} y + \frac{1}{z} \frac{\partial}{\partial z} y + \left(1 - \frac{\nu^2}{z^2} \right) y = 0$$

where $\nu^2 = -2E$.

P3.3. Verify the conservation law (3.1c) and show that the Hamiltonian H of (3.1d) is a constant of motion.

P3.4. Show that (3.7) indicates that $\int_0^{2\pi} d\varphi \int_0^{\pi} \sin\theta \, d\theta \int_0^{\infty} |E|^2 r^2 \, dr$ is conserved.

P3.5. Find the plane wave solutions of the nonlinear Schrödinger equation

$$i\psi_t + \psi_{xx} + \alpha|\psi|^2\psi = 0 \tag{P3.2}$$

(1) With the initial condition,

$$\psi(x,0) = C_1 \exp[i(C_2 x + C_3)]$$

(2) With the boundary condition,

$$\psi(0,t) = \frac{C_1}{\sqrt{t}} \exp\left\{i\left[\frac{C_2^2}{4t} + (\alpha C_1^2 \ln t + C_3)\right]\right\}$$

where $C_i, i = 1, 2, 3, \ldots$ are constants.

P3.6. Find the solitary solutions of the nonlinear Schrödinger equation (P3.2) with the initial conditions

(1) $\psi(x,0) = \pm C_1\sqrt{\frac{2}{\alpha}} \exp(iC_2) \operatorname{sech}(C_1 x + C_3)$ for a stationary soliton.

(2) $\psi(x,0) = \pm A\sqrt{\frac{2}{\alpha}} \exp[i(Bx + C_1)] \operatorname{sech}(Ax + C_2)$ for a propagating soliton.

P3.7. Find the solitary solution, i.e., (3.5b), of (3.2) via an associated Legendre polynomial.

Introduce a transform, $s = \tanh k\xi$, on (3.2), which maps ξ from $(-\infty, \infty)$ to s from $(-1, 1)$.

(1) Show that (3.2) becomes

$$\frac{d}{ds}(1 - s^2)\frac{d}{ds}\phi + 2\frac{\alpha\phi^2 + E}{k^2(1 - s^2)}\phi = 0 \tag{P3.3}$$

(2) Determine ϕ, with the aid of the similarity of the equation form to the associated Legendre differential equation

$$\frac{d}{ds}(1 - s^2)\frac{d}{ds}P_N^n + [N(N + 1) - \frac{n^2}{(1 - s^2)}]P_N^n = 0 \tag{P3.3a}$$

where

$$P_N^n(s) = \frac{(-1)^n}{2^N N!}(1 - s^2)^{\frac{n}{2}}\frac{d^{N+n}}{ds^{N+n}}(s^2 - 1)^N \tag{P3.3b}$$

It is noted that $P_1^1 = -\frac{1}{2}(1 - s^2)^{\frac{1}{2}}\frac{d^2}{ds^2}(s^2 - 1) = -(1 - s^2)^{\frac{1}{2}} = \mp \operatorname{sech} k\xi$].

P3.8. With the aid of (1.25b),

   (1) decouple the sets of coupled equations (1.26) and (1.27);

   (2) convert (1.26) and (1.27) into the equations for ϕ and ϕ^* which separate the spatial and time derivatives;

   (3) show that (1.32a) is a solution of the equation for ϕ, derived in (2);

   (4) convert the decoupled equations of (1.26), derived in (1), to the form of (P3.1).

P3.9. Integrate the mKdV equation (3.15a) for a propagating mode to obtain a solitary solution.

P3.10. Find the solutions of the Burgers equation (1.8a) for the following initial conditions of the diffusion equation (1.8e) with the aid of the Bäcklund transform (1.8c):

   (1) $v(x,0) = \delta(x)$,

   (2) $v(x,0) = 1 + \delta(x)$,

   (3) $v(x,0) = V_0 \sin \frac{\pi x}{L}$.

Chapter 4

Inverse Scattering Transform (IST)

In a quantum mechanical system, the potential function could be reconstructed via the scattering data. In other words, a localized potential field $V(x)$ of the linear Schrödinger equation

$$\varphi_{xx} + [E - V(x)]\varphi = 0 \tag{4.1a}$$

can be completely reconstructed from the scattering data S, which include the reflection and transmission coefficients of a spectral wave in unbound states and the eigenvalues and the asymptotic coefficients of the normalized eigenfunctions in bound states. It is noted that (4.1a) also represents a linear inhomogeneous Helmholtz equation describing the propagation of time-harmonic waves in an inhomogeneous dielectric medium.

Set $E = k^2$ and take the inverse Fourier transform of (4.1a), it becomes

$$\left[\frac{\partial^2}{\partial x^2} - \frac{\partial^2}{\partial y^2} - V(x)\right]\hat{\varphi}(x, y) = 0 \tag{4.1b}$$

where

$$\hat{\varphi}(x, y) = \int_{-\infty}^{\infty} \varphi(x, k)e^{iky}\frac{dk}{2\pi}$$

It shows that $\hat{\varphi}(x, y)$ is a solution of a linear wave equation and $V(x)$ represents a wave scatterer. If $V(x)$ is given, (4.1a) can be solved to obtain the scattering data. On the other hand, an unknown scatterer $V(x)$ of (4.1a) may be reconstructed through a set of known scattering data. A procedure of this inverse scattering transform to determine $V(x)$ from the scattering data is described in the following.

Exercise 4.1: Show that an inverse Fourier transform converts (4.1a) into (4.1b).

4.1   Scattering Problem

Consider the propagation of a wave, represented by a phasor function e^{-ikx}, along the x-axis from $x = \infty$ to $x = -\infty$ (i.e., from right to left). If $V(x;t)$ is in the form of a localized potential well, it causes reflection and transmission of the wave, as well as traps wave in the potential well. Then, there will be a finite number of bound states with discrete eigenenergies $E_n = -\kappa_n^2; n = 1, 2, \ldots, N$, and a continuum of states with eigenenergies $E = k^2$. These eigenvalues determine the corresponding eigenfunctions asymptotically as follows:

1. For the bound states

$$\varphi_n \sim c_n e^{-\kappa_n x} \text{ as } x \to +\infty \qquad (4.2a)$$

and

$$\varphi_n \sim e^{\kappa_n x} \text{ as } x \to -\infty$$

where the scattering amplitude c_n is determined via the normalization $\int_{-\infty}^{\infty} \varphi_n^2 \, dx = 1$.
2. For an unbound state

$$\varphi \sim e^{-ikx} + b(k)e^{ikx} \text{ as } x \to \infty \qquad (4.2b)$$

and

$$\varphi \sim a(k)e^{-ikx} \text{ as } x \to -\infty$$

where $b(k)$ represents a reflection coefficient and $a(k)$ is a transmission coefficient.

These asymptotic results render the scattering data $S = \{\kappa_n, c_n; a(k), b(k)\}$ and a representation of the Jost solution for

(4.1a) as

$$\varphi(x, \lambda) = \varphi_0(x, \lambda) + \int_x^\infty K(x, z)\varphi_0(z, \lambda)\, dz \qquad (4.2c)$$

where, for $x > 0$,

$$\varphi_0^+(x, \lambda = k) = e^{-ikx} + b(k)e^{ikx} \text{ for unbound states}$$
$$\varphi_{n0}^+(x, \lambda = i\kappa_n) = c_n e^{-\kappa_n x} \text{ for bound states}$$

and the kernel $K(x, z)$ of the integral is the solution of the Gelfand–Levitan–Marchenko (GLM) linear integral equation.

4.2 Gelfand–Levitan–Marchenko (GLM) Linear Integral Equation

Define the function associated with the backward scattering (toward to $+x$ direction) data as

$$F(x) = \sum_{n=1}^{N} c_n^2 e^{-\kappa_n x} + \frac{1}{2\pi} \int_{-\infty}^{\infty} b(k)e^{ikx}\, dk \qquad (4.3)$$

then, the potential $V(x)$ of the linear Schrödinger equation (4.1a), responsible for the backward scattering, is restored from the formula

$$V(x) = -2\frac{\partial}{\partial x}K(x, y = x) \qquad (4.4)$$

where $K(x, y)$ is found from the GLM linear integral equation

$$K(x, y) + F(x + y) + \int_x^\infty K(x, z)F(y + z)\, dz = 0 \qquad (4.5)$$

That is, from the scattering data S of (4.1a), the corresponding inverse scattering transform (IST) mapping $S \to V$ is accomplished through solving the GLM linear integral equation. Derivations of (4.4), (4.5), and (4.3) are presented in detail in Sections 5.1 and 5.2 of Chapter 5.

It can be shown that the wave function (4.2c) satisfies the linear Schrödinger equation (4.1a) if the kernel $K(x, z)$ is the solution of the GLM equation (4.5) and the potential function $V(x)$ of (4.1a) is given by (4.4). Proof is presented in Section 5.3 of Chapter 5.

For a time-dependent potential function $V(x, t)$, "t" is treated as a parameter in the IST; specifically, $V(x; t_0) = V(x, t_0)$ determines a set of scattering data $S(t_0) = \{\kappa_n, c_n(t_0); a(k; t_0), b(k; t_0)\}$ through (4.1a); likewise, $V(x; t_0)$ can be reconstructed, with the aid of $S(t_0)$, by the IST, i.e., the functions in (4.3) to (4.5) become $F(x; t_0), K(x, y; t_0), V(x; t_0), c_n(t_0)$ and $b(k; t_0)$. Therefore, if the time-dependent scattering data is known, i.e., $S = S(t)$, then IST reconstructs $V(x, t) = -2\frac{\partial}{\partial x} K(x, y = x; t)$.

It suggests that the inverse scattering transform may be applied for solving some nonlinear partial differential equations (NLPDEs), by embedding the solution $\phi(x, t)$ of the NLPDE as the potential function $V(x; t)$, e.g., $\phi(x, t) = -V(x; t)$, of (4.1a). For an initial value problem, $\phi(x, 0)$ is given, then $V(x) = -\phi(x, 0)$ can be employed in (4.1a) to evaluate the initial scattering data $S(0)$. However, the information on the time evolution of scattering data, i.e., $S(t)$, is necessary to reconstruct $V(x; t)$ via IST.

Equation (4.1a) is an auxiliary equation of the IST method for solving an NLPDE; to ensure that $\phi(x, t) = -V(x; t)$ is the solution of this NLPDE, a second auxiliary equation of $\varphi(x; t)$, a rate equation, needs to be found; the two coupled auxiliary equations set a non-trivial solution condition that requires the potential function $\phi(x; t)$ to satisfy the NLPDE. It turns out that this rate equation (the second auxiliary equation) works to update the scattering data in time, from $S(0)$ to $S(t)$. The two auxiliary equations can be converted into operator forms or matrix forms. The two operators or matrices, forming "a Lax pair", are governed by a Lax equation, which represents this NLPDE. On the other hand, the Lax equation guides us to find the Lax pair of the NLPDE.

It is noted that the potential function $V(x)$ of (4.1a) may represent a more complex function associated with $\phi(x, 0)$ and its spatial derivatives, e.g., $V(x) = F[\phi(x, 0), \phi_x(x, 0), \ldots]$; in this case, IST reconstructs $F[\phi(x, t), \phi_x(x, t), \ldots]$, which is then solved to obtain $\phi(x, t)$. However, solving this equation $F[\phi(x, t), \phi_x(x, t), \ldots] = V(x, t)$ set up via the IST may not be straightforward. For instance, Miura transformation (3.14) converts the KdV equation (1.1) into the mKdV equation (3.15a); it also transforms the two auxiliary equations of the KdV equations to the auxiliary equations of the mKdV equation. Thus, the potential function of (4.1a) becomes $V(x) = F[U(x,0), U_x(x,0)] = U^2(x,0) + U_x(x,0)$. In this approach, IST reconstructs $U^2(x,t) + U_x(x,t)$, rather than the solution $U(x,t)$ directly, of the mKdV equation.

4.3   Nonlinear Synthesis

4.3.1   *Auxiliary equations*

Auxiliary equations synthesizing a nonlinear partial differential equation, which has a solution $\phi(x,t)$, are introduced, and one of those is in the form of a linear Schrödinger equation. It is achieved by introducing an auxiliary function $\psi(x,t;\lambda)$ which satisfies a linear Schrödinger equation

$$\psi_{xx} = -(\phi + \lambda)\psi \qquad (4.6)$$

where λ is a constant and $-\phi(x,t)$ represents the potential function $V(x)$ of this quantum system with t treated as a parameter, i.e., ϕ and ψ depend parametrically on t. It is noted that (4.6) and (4.1a) are essentially the same equations. For an initial value problem, the initial condition $\phi(x,0)$ determines the initial scattering data $S(0) = \{\kappa_n, c_n(0); a(k;0), b(k;0)\}$. On the other hand, the time evolution of the scattering data $S(t) = \{\kappa_n, c_n(t); a(k;t), b(k;t)\}$ needs to be determined to proceed IST. Moreover, a condition must be imposed to ensure the potential function $\phi(x,t)$ satisfying this nonlinear partial differential equation. Thus, a second auxiliary equation of $\psi(x,t;\lambda)$, in a rate equation form, is introduced; two coupled auxiliary equations set a non-trivial solution condition, manifesting that $\phi(x,t)$ is a solution of this nonlinear partial differential equation. In other words, this pair of auxiliary equations represents the original NLPDE. A derivation of the second auxiliary equation and the necessity of this equation are presented next, exemplified through the KdV equation

$$\phi_t + 6\phi\phi_x + \phi_{xxx} = 0 \qquad (4.7)$$

We now consider the solution $\phi(x, t)$ of the KdV equation (4.7) to be a potential function of a quantum system (4.6) (or a dielectric function of a medium), where t is treated as a parameter, i.e., $\phi(x, t) = \phi(x; t)$, then the problem of solving a nonlinear partial differential equation (the KdV equation) with x and t variables is first reduced to solving a linear inhomogeneous Schrödinger (Helmholtz) equation (an ordinary differential equation with x variable) to determine the scattering data; Although, at this stage, $\phi(x, t)$ is yet unknown, the initial condition $\phi(x, 0)$ of an initial value problem is available to determine the initial scattering data $S(0)$. But the information $S(0)$ alone is not adequate to reconstruct $\phi(x; t)$. Moreover, the auxiliary equation (4.6) alone does not represent the KdV equation. Thus, a second auxiliary equation of $\psi(x, t; \lambda)$ is introduced, which, together with (4.6), can represent the KdV equation. Since the initial scattering data $S(0)$ needs to update in time, the second auxiliary equation must be in a rate equation form

$$\psi_t = \vec{A}\psi \qquad (4.8)$$

where $\vec{A} = F(\phi, \phi_x) + G(\phi, \phi_x)\partial_x$ is a differential operator and functions F and G are determined in the following.

From (4.6) and (4.8), one can obtain ψ_{xxt} and, ψ_{txx}. Impose KdV equation on ϕ, the condition of $\psi_{xxt} = \psi_{txx}$ set equations in the following to determine F and G. With the aid of (4.6),

$$\psi_{xxt} = -\phi_t\psi - (\phi + \lambda)\psi_t = (\phi_{xxx} + 6\phi\phi_x)\psi - (\phi + \lambda)\vec{A}\psi$$

With the aid of (4.8),

$$\psi_{txx} = \partial_x^2 \vec{A}\psi = \vec{A}\psi_{xx} + 2(F_x + G_x\partial_x)\psi_x + (F_{xx} + G_{xx}\partial_x)\psi$$

where

$$\vec{A}\psi_{xx} = -\vec{A}(\phi + \lambda)\psi = -(\phi + \lambda)\vec{A}\psi - G\phi_x\psi$$

Thus, the relation $\psi_{xxt} = \psi_{txx}$ sets up

$$-G\phi_x\psi + 2(F_x + G_x\partial_x)\psi_x + (F_{xx} + G_{xx}\partial_x)\psi = (\phi_{xxx} + 6\phi\phi_x)\psi$$

The LHS is rearranged as

$$[F_{xx} - 2G_x(\phi + \lambda) - G\phi_x]\psi + (2F_x + G_{xx})\psi_x = (\phi_{xxx} + 6\phi\phi_x)\psi$$

and it is deduced as the two equations

$$2F_x + G_{xx} = 0$$

and

$$F_{xx} - 2G_x(\phi + \lambda) - G\phi_x = (\phi_{xxx} + 6\phi\phi_x)$$

Set $F_{xx} = \phi_{xxx}$, it results to

$$F = \phi_x + \gamma \text{ and } G = -2\phi + 4\lambda$$

Thus, (4.8) becomes

$$\psi_t = (\phi_x + \gamma)\psi + (4\lambda - 2\phi)\psi_x \tag{4.8a}$$

The auxiliary equations (4.6) and (4.8a) synthesize the KdV equation. Equation (4.6) defines the scattering problem and (4.8a) defines the time evolution of the scattering.

4.3.2 *Lax pair and Lax equation*

4.3.2.1 *Operator form*

Introduce a differential operator

$$\vec{L} = -\frac{\partial^2}{\partial x^2} - \phi \tag{4.6a}$$

Then, with the aid of (4.6a), (4.6) is re-expressed as

$$\vec{L}\psi = \lambda\psi \tag{4.6b}$$

Take $\frac{\partial}{\partial t}$ of (4.6b) and the aid of (4.8), it yields

$$\frac{\partial}{\partial t}\vec{L}\psi = \vec{L}_t\psi + \vec{L}\psi_t = \vec{L}_t\psi + \vec{L}\vec{A}\psi = \lambda_t\psi + \lambda\psi_t = \vec{A}\lambda\psi = \vec{A}\vec{L}\psi$$

where $\lambda_t = 0$ is set, i.e., λ is independent of time; then, it leads to the Lax equation

$$\vec{L}_t + [\vec{L}, \vec{A}] = 0 \tag{4.9}$$

where $[\vec{L}, \vec{A}] = \vec{L}\vec{A} - \vec{A}\vec{L}$ represents operator commutator.

With the aid of (4.6), the parameter λ in (4.8a) can be expressed in terms of differential operators. Replace $\lambda\psi_x$ by $\frac{\partial}{\partial x}\vec{L}\psi = -\frac{\partial}{\partial x}(\frac{\partial^2}{\partial x^2} + \phi)\psi$, (4.8a) becomes

$$\psi_t = (\phi_x + \gamma)\psi - 2\phi\frac{\partial}{\partial x}\psi - 4\left(\frac{\partial^3}{\partial x^3} + \phi\frac{\partial}{\partial x} + \phi_x\right)\psi$$

$$= \left(\gamma - 4\frac{\partial^3}{\partial x^3} - 6\phi\frac{\partial}{\partial x} - 3\phi_x\right)\psi \qquad (4.8b)$$

Thus,

$$\vec{A} = \gamma - 4\frac{\partial^3}{\partial x^3} - 6\phi\frac{\partial}{\partial x} - 3\phi_x \qquad (4.8c)$$

The two operators $\vec{L}$ and $\vec{A}$ form a Lax pair. The Lax equation (4.9) represents a given PDE, subject to a specified Lax pair. For instance, if the Lax pair $\vec{L}$ and $\vec{A}$ is specified by (4.6a) and (4.8c), respectively, then, (4.9) represents the KdV equation. It is verified as follows.

Take time derivative on (4.6a), it gives

$$\vec{L}_t = -\phi_t$$

Next, perform the operator commutator $[\vec{L}, \vec{A}]$ with $\vec{L}$ and $\vec{A}$ given by (4.6a) and (4.8c), respectively; it gives

$$[\vec{L}, \vec{A}] = \vec{L}\vec{A} - \vec{A}\vec{L}$$

$$= -\left(\frac{\partial^2}{\partial x^2} + \phi\right)\left(\gamma - 4\frac{\partial^3}{\partial x^3} - 6\phi\frac{\partial}{\partial x} - 3\phi_x\right)$$

$$+ \left(\gamma - 4\frac{\partial^3}{\partial x^3} - 6\phi\frac{\partial}{\partial x} - 3\phi_x\right)\left(\frac{\partial^2}{\partial x^2} + \phi\right)$$

$$= \left[\frac{\partial^2}{\partial x^2}\left(6\phi\frac{\partial}{\partial x} + 3\phi_x\right) - \left(6\phi\frac{\partial}{\partial x} + 3\phi_x\right)\frac{\partial^2}{\partial x^2}\right]$$

$$+ \left[\phi\left(4\frac{\partial^3}{\partial x^3} + 6\phi\frac{\partial}{\partial x}\right) - \left(4\frac{\partial^3}{\partial x^3} + 6\phi\frac{\partial}{\partial x}\right)\phi\right]$$

With the aid of

$$\frac{\partial^2}{\partial x^2}\left(\phi\frac{\partial}{\partial x}\right) = \phi_{xx}\frac{\partial}{\partial x} + 2\phi_x\frac{\partial^2}{\partial x^2} + \phi\frac{\partial^3}{\partial x^3}$$

$$\frac{\partial^2}{\partial x^2}\phi_x = \phi_x\frac{\partial^2}{\partial x^2} + 2\phi_{xx}\frac{\partial}{\partial x} + \phi_{xxx}$$

$$\frac{\partial^3}{\partial x^3}\phi = \phi\frac{\partial^3}{\partial x^3} + 3\phi_x\frac{\partial^2}{\partial x^2} + 3\phi_{xx}\frac{\partial}{\partial x} + \phi_{xxx}$$

$$\frac{\partial}{\partial x}\phi = \phi_x + \phi\frac{\partial}{\partial x}$$

The commutator becomes

$$[\vec{L},\,\vec{A}] = -6\phi\phi_x - \phi_{xxx}$$

Then, the Lax equation (4.9) leads to the KdV equation

$$\phi_t + 6\phi\phi_x + \phi_{xxx} = 0$$

On the other hand, the Lax equation also works, with the aid of the NLPDE to be solved, to find the operator $\vec{A}$ for a likely $\vec{L}$ of the NLPDE. Again, it is exemplified, in the following, through the KdV equation (4.7); so, the operator $\vec{L}$ is given by (4.6a). Assume that

$$\vec{A} = a_0 + a_1\frac{\partial}{\partial x} + a_2\frac{\partial^2}{\partial x^2} + a_3\frac{\partial^3}{\partial x^3}$$

Then,

$$[\vec{L},\,\vec{A}]$$

$$= -\left(\frac{\partial^2}{\partial x^2} + \phi\right)\left(a_0 + a_1\frac{\partial}{\partial x} + a_2\frac{\partial^2}{\partial x^2} + a_3\frac{\partial^3}{\partial x^3}\right)$$

$$+ \left(a_0 + a_1\frac{\partial}{\partial x} + a_2\frac{\partial^2}{\partial x^2} + a_3\frac{\partial^3}{\partial x^3}\right)\left(\frac{\partial^2}{\partial x^2} + \phi\right)$$

With the aid of

$$\frac{\partial^2}{\partial x^2}\left(a_0 + a_1\frac{\partial}{\partial x} + a_2\frac{\partial^2}{\partial x^2} + a_3\frac{\partial^3}{\partial x^3}\right)$$

$$= \left(a_0 + a_1\frac{\partial}{\partial x} + a_2\frac{\partial^2}{\partial x^2} + a_3\frac{\partial^3}{\partial x^3}\right)\frac{\partial^2}{\partial x^2}$$

$$+ 2\left(a_{0_x} + a_{1_x}\frac{\partial}{\partial x} + a_{2_x}\frac{\partial^2}{\partial x^2} + a_{3_x}\frac{\partial^3}{\partial x^3}\right)\frac{\partial}{\partial x}$$

$$+ \left(a_{0_{xx}} + a_{1_{xx}}\frac{\partial}{\partial x} + a_{2_{xx}}\frac{\partial^2}{\partial x^2} + a_{3_{xx}}\frac{\partial^3}{\partial x^3}\right)$$

and

$$\left(a_0 + a_1\frac{\partial}{\partial x} + a_2\frac{\partial^2}{\partial x^2} + a_3\frac{\partial^3}{\partial x^3}\right)\phi$$

$$= \phi\left(a_0 + a_1\frac{\partial}{\partial x} + a_2\frac{\partial^2}{\partial x^2} + a_3\frac{\partial^3}{\partial x^3}\right) + a_1\phi_x$$

$$+ a_2\left(\phi_{xx} + 2\phi_x\frac{\partial}{\partial x}\right)$$

$$+ \left(a_3\phi_{xxx} + 3\phi_{xx}\frac{\partial}{\partial x} + 3\phi_x\frac{\partial^2}{\partial x^2}\right)$$

The commutator becomes

$$[\vec{L}, \vec{A}] = -[a_{0_{xx}} - (a_1\phi_x + a_2\phi_{xx} + a_3\phi_{xxx})]$$

$$- [(2a_{0_x} + a_{1_{xx}}) - (2a_2\phi_x + 3a_3\phi_{xx})]\frac{\partial}{\partial x}$$

$$- [(2a_{1_x} + a_{2_{xx}}) - 3a_3\phi_x]\frac{\partial^2}{\partial x^2}$$

$$- (2a_{2_x} + a_{3_{xx}})\frac{\partial^3}{\partial x^3} - 2a_{3_x}\frac{\partial^4}{\partial x^4}$$

Applying the Lax equation and the KdV equation ($-\phi_t = 6\phi\phi_x + \phi_{xxx}$), the following relations are obtained:

$$(1)\, a_{3_x} = 0, \quad (2)\, 2a_{2_x} + a_{3_{xx}} = 0, \quad (3)\, 2a_{1_x} + a_{2_{xx}} = 3a_3\phi_x,$$

$$(4)\, 2a_{0_x} + a_{1_{xx}} = 2a_2\phi_x + 3a_3\phi_{xx}, \text{ and}$$

$$(5)\, a_{0_{xx}} - (a_1\phi_x + a_2\phi_{xx} + a_3\phi_{xxx}) = 6\phi\phi_x + \phi_{xxx}$$

The first two relations imply that $a_2 = 0$ and $a_3 = $ const.; the next two relations lead to $a_{1_x} = \frac{3}{2}a_3\phi_x$ and $a_{0_x} = \frac{3}{4}a_3\phi_{xx}$; then, the last relation leads to $a_3 = -4$, $a_1 = -6\phi$, and $a_0 = -3\phi_x + \gamma$, where γ is an integration constant. Thus,

$$\vec{A} = \gamma - 3\phi_x - 6\phi\frac{\partial}{\partial x} - 4\frac{\partial^3}{\partial x^3}$$

which is (4.8c).

Exercise 4.2: Show that (4.8a) converts to (4.8b).

4.3.2.2 *Matrix form*

In the operator form, $\vec{L}$ is a second-order differential operator and ψ is a scalar auxiliary function. If ψ is set to be a vector auxiliary function, i.e., a 1×2 row matrix $[\psi] = \begin{bmatrix} \Theta \\ \Psi \end{bmatrix}$ then, $\vec{L}$ is replaced by a 2×2 matrix $[L] = \begin{bmatrix} l_{11} & l_{12} \\ l_{21} & l_{22} \end{bmatrix}$, in which Θ and Ψ are coupled functions and the matrix elements l_{11} and l_{22} are first-order spatial differential operators, i.e., $l_{11} \propto \partial_x \propto l_{22}$. Thus, a matrix form of (4.6b) is

$$[L][\psi] = \lambda[\psi] \tag{4.6c}$$

Accordingly, (4.8) has the matrix form

$$[\psi]_t = [A][\psi] \tag{4.8d}$$

where $[A] = \begin{bmatrix} a_{11} & a_{12} \\ a_{21} & a_{22} \end{bmatrix}$.

Take $\frac{\partial}{\partial t}$ of (4.6c) and the aid of (4.8d) and set $\lambda_t = 0$, the Lax equation for matrix operators is derived as

$$[L]_t + \{[L], [A]\} = 0 \tag{4.9a}$$

where $\{[L], [A]\} = [L][A] - [A][L]$ is defined as the matrix commutator.

It is noted that (4.9c) has a similar form as (4.9), besides that the differential operators $\vec{L}$ and $\vec{A}$ are expressed in the matrix forms $[L]$ and $[A]$.

With regard to the KdV equation (4.7), $l_{11} = i\partial_x = -l_{22}, l_{12} = -i\phi$, and $l_{21} = -i$ are set in $[L]$ to give

$$[L] = i \begin{bmatrix} \partial_x & -\phi \\ -1 & -\partial_x \end{bmatrix} \tag{4.6d}$$

so that the determinant of the matrix $[L] - [\lambda]$ equals to $\partial_x^2 + \lambda^2 + \phi$, like the operator acting on ψ in (4.6); then,

$$[L]_t = i \begin{bmatrix} 0 & -\phi_t \\ 0 & 0 \end{bmatrix} = i \begin{bmatrix} 0 & 6\phi\phi_x + \phi_{xxx} \\ 0 & 0 \end{bmatrix}$$

In $[A]$, $a_{11} = a_{22}$ and $a_{21} = 0$ are set to give $[A] = \begin{bmatrix} a_{11} & a_{12} \\ 0 & a_{11} \end{bmatrix}$; then,

$[L][A] - [A][L]$

$$= i \begin{bmatrix} (\partial_x a_{11} - a_{11}\partial_x) + a_{12}(\partial_x a_{12} + a_{12}\partial_x) - (\phi a_{11} - a_{11}\phi) \\ 0 \qquad\qquad\qquad\qquad -(\partial_x a_{11} - a_{11}\partial_x) - a_{12} \end{bmatrix}$$

The Lax equation $[L]_t + \{[L], [A]\} = 0$ sets up two equations for the two unknowns a_{11} and a_{12}:

$$(\partial_x a_{11} - a_{11}\partial_x) + a_{12} = 0$$

and

$$(\partial_x a_{12} + a_{12}\partial_x) - (\phi a_{11} - a_{11}\phi) = -(6\phi\phi_x + \phi_{xxx})$$

Set

$$a_{11} = a_0 + a_1\partial_x + a_2\partial_x^2 + a_3\partial_x^3 \text{ and } a_{12} = b_0 + b_1\partial_x$$

in the two equations and apply the operations

$$\partial_x a_{11} = a_{11}\partial_x + (a_{0_x} + a_{1_x}\partial_x + a_{2_x}\partial_x^2 + a_{3_x}\partial_x^3)$$
$$\partial_x a_{12} = a_{12}\partial_x + (b_{0_x} + b_{1_x}\partial_x)$$
$$a_{11}\phi = \phi a_{11} + [a_1\phi_x + a_2(\phi_{xx} + 2\phi_x\partial_x)$$
$$+ a_3(\phi_{xxx} + 3\phi_{xx}\partial_x + 3\phi_x\partial_x^2)]$$

then, use the order of the differential operator "∂_x" to group terms; the two equations become

$$(a_{0_x} + b_0) + (a_{1_x} + b_1)\partial_x + a_{2_x}\partial_x^2 + a_{3_x}\partial_x^3 = 0$$

and

$$(b_{0_x} + a_1\phi_x + a_2\phi_{xx} + a_3\phi_{xxx}) + (2b_0 + b_{1_x} + 2a_2\phi_x + 3a_3\phi_{xx})\partial_x$$
$$+ (2b_1 + 3a_3\phi_x)\partial_x^2 = -(6\phi\phi_x + \phi_{xxx})$$

which imply the following relations:

$$(1)\, a_{2_x} = 0 = a_{3_x}, \quad (2)\, a_{1_x} + b_1 = 0, \quad (3)\, a_{0_x} + b_0 = 0,$$
$$(4)\, 2b_1 + 3a_3\phi_x = 0, \quad (5)\, 2b_0 + b_{1_x} + 2a_2\phi_x + 3a_3\phi_{xx} = 0, \text{ and}$$
$$(6)\, b_{0_x} + a_1\phi_x + a_2\phi_{xx} + a_3\phi_{xxx} = -(6\phi\phi_x + \phi_{xxx})$$

The first relation leads to $a_2 = 0$ and $a_3 = $ const.; relations (4) and (2) imply that $b_1 \propto \phi_x$ and $a_1 \propto \phi$, and relation (6) suggests that $a_1 = -6\phi$. Then, (2), (4), and (5) lead to $b_1 = 6\phi_x, a_3 = -4$, and $b_0 = 3\phi_{xx}$ and (6) is matched. Finally, $a_0 = -3\phi_x + \gamma$ is determined by (3). The results give

$$a_{11} = \gamma - 4\partial_x^3 - 6\phi\partial_x - 3\phi_x \text{ and } a_{12} = 6\phi_x\partial_x + 3\phi_{xx}$$

Then,

$$[A] = \begin{bmatrix} \gamma - 4\partial_x^3 - 6\phi\partial_x - 3\phi_x & 6\phi_x\partial_x + 3\phi_{xx} \\ 0 & \gamma - 4\partial_x^3 - 6\phi\partial_x - 3\phi_x \end{bmatrix} \quad (4.8e)$$

With the matrix form Lax pair $[L]$ and $[A]$ defined by (4.6d) and (4.8e), respectively, (4.9a) represents the KdV equation. Equations (4.6c) and (4.8d) also reproduce (4.6) and (4.8b), respectively, with ψ replaced by Ψ.

Exercise 4.3: Show that the matrix Lax equation (4.9a), with the Lax pair (4.6d) and (4.8e), represents the KdV equation.

Exercise 4.4: Show that (4.6c) and (4.8d) reproduce (4.6) and (4.8b), respectively, with ψ replaced by Ψ.

4.3.3   *Matrix AKNS pair and AKNS equation*

Ablowitz, Kaup, Newell, and Segur introduce another method to obtain a matrix form operator equation compatible with (4.9a). They replace (4.6c) by

$$[\psi]_x = [X][\psi] \tag{4.10}$$

and keep (4.8d), but using the notation $[T]$ to replace $[A]$, i.e.,

$$[\psi]_t = [T][\psi] \tag{4.11}$$

where $[X] = \begin{bmatrix} X_{11} & X_{12} \\ X_{21} & X_{22} \end{bmatrix}$ and $[T] = \begin{bmatrix} T_{11} & T_{12} \\ T_{21} & T_{22} \end{bmatrix}$; the matrix elements X_{ij} and T_{ij} do not comprise spatial differential operators.

With the aid of $[\psi]_{xt} = [\psi]_{tx}$, (4.10) and (4.11) lead to

$$[\psi]_{xt} = [X]_t[\psi] + [X][\psi]_t = [X]_t[\psi] + [X][T][\psi]$$
$$[\psi]_{tx} = [T]_x[\psi] + [T][\psi]_x = [T]_x[\psi] + [T][X][\psi]$$

It implies

$$[X]_t - [T]_x + \{[X], [T]\} = 0 \tag{4.12}$$

This is the matrix form of the AKNS equation, and $[X]$ and $[T]$ form an AKNS pair.

Again, consider the KdV equation (4.7); set $X_{11} = 0 = X_{22}, X_{12} = \lambda + \phi$, and $X_{21} = -1$ to give

$$[X] = \begin{bmatrix} 0 & \lambda + \phi \\ -1 & 0 \end{bmatrix} \tag{4.10a}$$

so that the determinant of the matrix $[I]\partial_x - [X]$ equals to $\partial_x^2 + \lambda + \phi$, also like the operator acting on ψ in (4.6), where $[I] = \begin{bmatrix} 1 & 0 \\ 0 & 1 \end{bmatrix}$, an identity matrix; then,

$$[X]_t = \begin{bmatrix} 0 & \phi_t \\ 0 & 0 \end{bmatrix} = \begin{bmatrix} 0 & -(6\phi\phi_x + \phi_{xxx}) \\ 0 & 0 \end{bmatrix}$$

In $[T]$, $T_{11} = -T_{22}$ is set to give $[T] = \begin{bmatrix} T_{11} & T_{12} \\ T_{21} & -T_{11} \end{bmatrix}$; then,

$$[T]_x = \begin{bmatrix} T_{11_x} & T_{12_x} \\ T_{21_x} & -T_{11_x} \end{bmatrix}$$

and

$$[X][T] - [T][X]$$
$$= \begin{bmatrix} (\lambda + \phi)T_{21} + T_{12} & -2(\lambda + \phi)T_{11} \\ -2T_{11} & -T_{21}(\lambda + \phi) - T_{12} \end{bmatrix}$$

The AKNS equation $[X]_t - [T]_x + \{[X], [T]\} = 0$ is applied to set up three equations for the unknowns $T_{11}, T_{12},$ and T_{21}:

$$T_{11_x} - [(\lambda + \phi)T_{21} + T_{12}] = 0$$
$$T_{21_x} + 2T_{11} = 0$$
$$T_{12_x} + 2(\lambda + \phi)T_{11} = -(6\phi\phi_x + \phi_{xxx})$$

Those equations are solved to give

$$T_{11} = -\phi_x, T_{12} = -\phi_{xx} - 2(\phi^2 - \lambda\phi - 2\lambda^2), \text{ and } T_{21} = 2(\phi - 2\lambda)$$

The matrix $[T]$ is then determined as

$$[T] = \begin{bmatrix} -\phi_x & -\phi_{xx} - 2(\phi^2 - \lambda\phi - 2\lambda^2) \\ 2(\phi - 2\lambda) & \phi_x \end{bmatrix} \qquad (4.11a)$$

With the AKNS pair $[X]$ and $[T]$ defined by (4.10a) and (4.11a), respectively, (4.12) represents the KdV equation. Likewise, (4.10) and (4.11) also reproduce (4.6) and (4.8b), respectively, with ψ replaced by Ψ.

Exercise 4.5: Show that the matrix AKNS equation (4.12), with the AKNS pair (4.10a) and (4.11a), represents the KdV equation.

Exercise 4.6: Show that (4.10) and (4.11) reproduce (4.6) and (4.8b), respectively, with ψ replaced by Ψ.

4.4   Time Evolution of Scattering Data

The initial scattering data $S(0) = \{\kappa_n, c_n(0); a(k;0), b(k;0)\}$ is updated in time through (4.8a).

1. For the bound states, (4.8a) gives

$$\frac{\psi_{nt}}{\psi_n} = \left(\phi_x - 2\phi \frac{\psi_{nx}}{\psi_n} \right) + \gamma_n + 4\lambda_n \frac{\psi_{nx}}{\psi_n}$$

where $\lambda_n = -\kappa_n^2$.
   As $x \to \infty$,

$$\frac{\psi_{nt}}{\psi_n} \to \frac{d}{dt} \ln c_n, \left(\phi_x - 2\phi \frac{\psi_{nx}}{\psi_n} \right) \to 0, \text{ and } \frac{\psi_{nx}}{\psi_n} \to -\kappa_n$$

thus,

$$\frac{d}{dt} \ln c_n = \gamma_n + 4\kappa_n^3 \tag{4.13}$$

In fact, (4.13) can be obtained directly from (4.8b), with the aid of (4.2a).

Exercise 4.7: Derive (4.13) via (4.8b) and (4.2a).

Ans: As $x \to +\infty$; $\phi \to 0$, $\varphi_n \sim c_n e^{-\kappa_n x}$, and $\varphi_{nt} \sim (\gamma_n - 4\frac{\partial^3}{\partial x^3})\varphi_n$.
   Since eigenfunctions are normalized to 1,

$$\frac{d}{dt} \int_{-\infty}^{\infty} \psi_n^2 dx = 0$$

With the aid of (4.8a),

$$\int_{-\infty}^{\infty} \psi_{nt}\psi_n dx = \int_{-\infty}^{\infty} [(\phi_x + \gamma_n)\psi_n + (4\lambda_n - 2\phi)\psi_{nx}]\psi_n dx$$

$$= \int_{-\infty}^{\infty} \{\gamma_n\psi_n^2 + [(\phi + 4\lambda_n)\psi_n^2]_x - 4(\phi + \lambda_n)\psi_n\psi_{nx}\}dx$$

$$= \int_{-\infty}^{\infty} \{\gamma_n\psi_n^2 + [(\phi + 4\lambda_n)\psi_n^2 + 2\psi_{nx}^2]_x\}dx = \gamma_n = 0$$

Then, (4.13) becomes

$$\frac{d}{dt}\ln c_n = 4\kappa_n^3$$

which is integrated as

$$c_n(t) = c_n(0)e^{4\kappa_n^3 t} \tag{4.14}$$

2. For the unbound states, Eq. (4.8a) gives

$$\frac{\psi^*\psi_t}{|\psi|^2} = \left(\phi_x - 2\phi\frac{\psi_x}{\psi}\right) + \gamma + 4\lambda\frac{\psi_x}{\psi}$$

$$\frac{\psi\psi_t^*}{|\psi|^2} = \left(\phi_x^* - 2\phi^*\frac{\psi_x^*}{\psi^*}\right) + \gamma^* + 4\lambda\frac{\psi_x^*}{\psi^*}$$

Thus,

$$\frac{d}{dt}\ln|\psi|^2 = \left(\phi_x - 2\phi\frac{\psi_x}{\psi}\right) + \gamma + 4\lambda\frac{\psi_x}{\psi} + \text{c.c.}$$

where $\lambda = k^2$.

(a) As $x \to -\infty$,

$$\frac{d}{dt} \ln |\psi|^2 \to \frac{d}{dt} \ln |a|^2, \left(\phi_x - 2\phi \frac{\psi_x}{\psi} \right) \to 0, \text{ and}$$

$$\frac{\psi_x}{\psi} \to -ik$$

thus,

$$\frac{d}{dt} \ln |a|^2 = \gamma + \gamma^* \qquad (4.15a)$$

(b) As $x \to \infty$,

$$\frac{d}{dt} \ln |\psi|^2 \to \frac{\frac{d}{dt}|b|^2 + \frac{d}{dt} be^{2ikx} + \frac{d}{dt} b^* e^{-2ikx}}{1 + |b|^2 + (be^{2ikx} + b^* e^{-2ikx})},$$

$$\left(\phi_x - 2\phi \frac{\psi_x}{\psi} \right) \to 0, \text{ and } \frac{\psi_x}{\psi} \to -ik \frac{1 - be^{2ikx}}{1 + be^{2ikx}}$$

thus,

$$\frac{\frac{d}{dt}|b|^2 + \frac{d}{dt} be^{2ikx} + \frac{d}{dt} b^* e^{-2ikx}}{1 + |b|^2 + (be^{2ikx} + b^* e^{-2ikx})}$$

$$= \gamma + \gamma^* - 4ik^3 \frac{1 - be^{2ikx}}{1 + be^{2ikx}} + 4ik^3 \frac{1 - b^* e^{-2ikx}}{1 + b^* e^{-2ikx}} \qquad (4.15b)$$

It implies that

$$\frac{d}{dt}|b|^2 = (\gamma + \gamma^*)(1 + |b|^2) \qquad (4.15c)$$

$$\frac{d}{dt} b = (\gamma + \gamma^* + 8ik^3)b \text{ and } \frac{d}{dt} b^*$$

$$= (\gamma + \gamma^* - 8ik^3)b^* \qquad (4.15d)$$

Since $|a|^2 + |b|^2 = 1$, (4.15a) and (4.15c) lead to $\gamma + \gamma^* = 0$; (4.15a) and (4.15d) are integrated to give

$$a(k;t) = a(k;0) \text{ and } b(k;t) = b(k;0)e^{8ik^3 t} \qquad (4.16)$$

In sum, with the aid of the second auxiliary equation (4.8a), the normalization of the eigenfunctions, the wave power conservation (the sum of the reflected power and transmitted power equals to the incident power), and the asymptotic properties of ψ (and ψ_n), the scattering data, to be used in the GLM integral equation, are updated. As a result, it is found that $b(k,t) = b(k,0)e^{8ik^3t}$, $a(k,t) = a(k,0)$, and $c_n(t) = c_n(0)e^{4\kappa_n^3 t}$. Thus,

$$S(t) = \{\kappa_n, \ c_n(0)e^{4\kappa_n^3 t}; \ a(k;0), \ b(k;0)e^{8ik^3t}\} \qquad (4.17)$$

Apply the scattering data (4.17) in (4.3), (4.5) is solved to obtain the kernel function $K(x,y;t)$, which reconstructs the potential field $V(x;t)$ of (4.1a) through the relation (4.4), and $\phi(x,t) = -V(x;t)$ is a solution of the KdV equation.

Exercise 4.8: Show that (4.15b) leads to (4.15c) and (4.15d).

4.5 Solving GLM Equation

Under a special class of initial conditions (having solitary wave solutions) which grant reflectionless potential in (4.6), i.e., $b(k;0) \sim 0$; then, (4.3) reduces to

$$F(x;t) = \sum_{n=1}^{N} c_n^2(t)e^{-\kappa_n x} \qquad (4.18)$$

GLM equation (4.5) is solved via the separation of variables by setting

$$K(x,y;t) = \sum_{n=1}^{N} H_n(x;t)e^{-\kappa_n y} \qquad (4.19)$$

Substitute (4.18) and (4.19) into (4.5), it becomes

$$\sum_{n=1}^{N} H_n(x;t)e^{-\kappa_n y} + \sum_{n=1}^{N} c_n^2(t)e^{-\kappa_n(x+y)}$$

$$+ \sum_{n=1}^{N} c_n^2(t)e^{-\kappa_n y} \sum_{n'=1}^{N} H_{n'}(x;t) \int_x^{\infty} e^{-(\kappa_{n'}+\kappa_n)z} \, dz = 0$$

$$(4.20a)$$

and infers that

$$H_n(x;t) + c_n^2(t)e^{-\kappa_n x} + c_n^2(t)\sum_{n'=1}^{N} H_{n'}(x;t)\frac{e^{-(\kappa_{n'}+\kappa_n)x}}{\kappa_{n'}+\kappa_n} = 0$$

$$(4.20b)$$

which is expressed in a matrix equation

$$(A_{ij})(H_j) = (C_i)$$

Then,

$$(H_i) = (A_{ij})^{-1}(C_j) = \frac{(B_{ij})}{|A_{ij}|}(C_j) \qquad (4.21)$$

where $i, j = 1, \ldots, N$; (A_{ij}) and (B_{ij}), and (H_i) and (C_i) are $N \times N$ and $1 \times N$ matrices, respectively; $(A_{ij})^{-1} = \frac{(B_{ij})}{|A_{ij}|}$ is the inversion of (A_{ij}); $|A_{ij}|$ is the determinant of (A_{ij});

$$A_{ij} = \delta_{ij} + c_i^2(t)\frac{e^{-(\kappa_i+\kappa_j)x}}{\kappa_i+\kappa_j} \text{ and } C_i = -c_i^2(t)e^{-\kappa_i x}$$

$$|A_{ij}| = \sum_{i=1}^{N} A_{i1}B_{1i} = \sum_{i=1}^{N} A_{1i}B_{i1}$$

$$B_{ij} = (-1)^{i+j}|\tilde{A}_{pq}|_{p\neq j,q\neq i}$$

$|\tilde{A}_{pq}|$ is the determinant of a $(N-1) \times (N-1)$ matrix;

$$\tilde{A}_{pq} = A_{pq} \text{ and } \delta_{ij} \text{ being the Kronecker delta.}$$

Hence,

$$K(x,x;t) = \frac{|A_{ij}|_x}{|A_{ij}|} = \frac{\partial}{\partial x}\ln|A_{ij}| \qquad (4.19a)$$

and

$$\phi(x;t) = 2\frac{\partial}{\partial x}K(x,x;t) = 2\frac{\partial^2}{\partial x^2}\ln|A_{ij}| \qquad (4.4a)$$

Exercise 4.9: Show that (4.20a) can infer (4.20b).

4.6 Evolution of an Impulse and a Rectangular Pulse in the KdV System (1.1)

As examples, the KdV equation (4.7) is solved via the IST method for the two initial conditions: (1) an impulse and (2) a rectangular pulse.

4.6.1 *Evolution of an impulse*

The initial condition is an impulse: $\phi(x,0) = A\delta(x)$, where $A > 0$. The linear Schrödinger equation (4.6) becomes

$$\psi_{xx} + \lambda\psi = 0, \quad \text{for } x \neq 0 \tag{4.22}$$

subject to a jump condition and a continuity condition at $x = 0$:

$$\psi_x(0^+) - \psi_x(0^-) = -A\psi(0) \text{ and } \psi(0^+) = \psi(0^-) \tag{4.23}$$

where $0^\pm = \pm\epsilon$ with $\epsilon \to 0$.

Depending on the eigenvalue $\lambda \lessgtr 0$, (4.22) is solved for the eigenfunctions of the bound states ($\lambda_n = -\kappa_n^2$) and of the unbound states ($\lambda = k^2$).

1. For the bound states, the eigenfunctions of (4.22) are written as

$$\psi_n = \begin{cases} c_n e^{-\kappa_n x}, & \text{for } x > 0 \\ c_n e^{\kappa_n x}, & \text{for } x < 0 \end{cases} \tag{4.24a}$$

where the scattering amplitude c_n is determined, via the normalization $\int_{-\infty}^{\infty} \psi_n^2 \, dx = 1$, as $c_n = \sqrt{\kappa_n}$. Apply the jump condition to (4.24a), it gives $\kappa_n = \frac{A}{2} = \kappa$, implying that there is only one bound state. It is noted that there is no bound state in the case of $A < 0$.

Exercise 4.10: Show that (4.24a) is an eigenfunction of (4.22) and meets the conditions (4.23).

2. For an unbound state, (4.22) has a general solution

$$\psi = \begin{cases} e^{-ikx} + b(k)e^{ikx}, & \text{for } x > 0 \\ a(k)e^{-ikx}, & \text{for } x < 0 \end{cases} \tag{4.24b}$$

where the reflection coefficient $b(k)$ and transmission coefficient $a(k)$ are related through the jump condition and continuity condition at

$x = 0$:

$$-ik(1 - b - a) = -A(1 + b)$$

and

$$1 + b = a$$

which are solved to obtain

$$b(k) = -\frac{A}{2ik + A} \text{ and } a(k) = \frac{2ik}{2ik + A}$$

Then, (4.3) is given as

$$F(x;t) = \kappa e^{-\kappa(x - 8\kappa^2 t)} + \frac{iA}{4\pi} \int_{-\infty}^{\infty} \frac{e^{ik(x + 8k^2 t)}}{k - iA/2} dk$$

Substitute it into (4.5) and set $K(x, y; t) = H_1(x; t)e^{-\kappa y}$, it results to

$$H_1(x;t) + \kappa \left[e^{-\kappa(x - 8\kappa^2 t)} + \frac{i}{2\pi} \int_{-\infty}^{\infty} \frac{e^{ik(x + 8k^2 t)}}{k - i\kappa} dk \right]$$

$$+ H_1(x;t) \left[\frac{1}{2} e^{-\kappa(2x - 8\kappa^2 t)} - \frac{\kappa}{2\pi} e^{-\kappa x} \int_{-\infty}^{\infty} \frac{e^{ik(x + 8k^2 t)}}{k^2 + \kappa^2} dk \right] = 0$$

where the exponential factor $e^{i(k - i\kappa)y}$ in the two integrals is set to one (i.e., $e^{i(k - i\kappa)y} = 1$) because both integrands have a pole at $k = i\kappa$. Then,

$$H_1(x;t) = -\kappa \frac{e^{-\kappa(x - 8\kappa^2 t)} + \frac{i}{2\pi} \int_{-\infty}^{\infty} \frac{e^{ik(x + 8k^2 t)}}{k - i\kappa} dk}{1 + \frac{1}{2} e^{-\kappa(2x - 8\kappa^2 t)} - e^{-\kappa x} \frac{\kappa}{2\pi} \int_{-\infty}^{\infty} \frac{e^{ik(x + 8k^2 t)}}{k^2 + \kappa^2} dk}$$

and

$$K^{(1)}(x,\,x;t) = -\kappa\frac{e^{-\kappa(2x-8\kappa^2 t)} + \frac{i}{2\pi}e^{-\kappa x}\int_{-\infty}^{\infty}\frac{e^{ik(x+8k^2 t)}}{k-i\kappa}\,dk}{1 + \frac{1}{2}e^{-\kappa(2x-8\kappa^2 t)} - e^{-\kappa x}\frac{\kappa}{2\pi}\int_{-\infty}^{\infty}\frac{e^{ik(x+8k^2 t)}}{k^2+\kappa^2}\,dk}$$

At $t = 0$ and for $x > 0$,

$$\int_{-\infty}^{\infty}\frac{e^{ikx}}{k^2+\kappa^2}\,dk = \oint\frac{e^{ikx}}{k^2+\kappa^2}\,dk = \frac{\pi}{\kappa}e^{-\kappa x}$$

where a semicircular arc of radius $R \to \infty$ is added in the upper half of the complex k-plane to convert line integral into contour integral (counterclockwise); thus,

$$K^{(1)}(x,\,x;0) = -\kappa\left[e^{-2\kappa x} + \frac{i}{2\pi}\int_{-\infty}^{\infty}\frac{e^{i(k+i\kappa)x}}{k-i\kappa}\,dk\right]$$

Then,

$$\phi(x,0) = 2\frac{\partial}{\partial x}K^{(1)}(x,\,x;0)$$

$$= 2\kappa\left[2\kappa e^{-2\kappa x} + \frac{1}{2\pi}\int_{-\infty}^{\infty}\frac{k+i\kappa}{k-i\kappa}e^{i(k+i\kappa)x}\,dk\right]$$

$$= A\delta(x)$$

it is consistent with the initial condition, where the integral is carried out as

$$\frac{1}{2\pi}\int_{-\infty}^{\infty}\frac{k+i\kappa}{k-i\kappa}e^{i(k+i\kappa)x}\,dk$$

$$= \frac{1}{2\pi}\int_{-\infty}^{\infty}e^{i(k+i\kappa)x}\,dk + \frac{i\kappa}{\pi}\oint\frac{e^{i(k+i\kappa)x}}{k-i\kappa}\,dk$$

$$= \delta(x) - 2\kappa e^{-2\kappa x}$$

For $t > 0$, $|e^{i8k^3t}| \to \infty$ as $\mathrm{Im}(k) \to \infty$; the integration along the semicircular arc of radius $R \to \infty$ in the upper half of the complex k-plane is infinite, thus the line integral cannot be converted into contour integral. On the other hand, a semicircular arc of radius $R \to \infty$ in the lower half of the complex k-plane can be added to convert the line integral in the denominator of $K^{(1)}(x, x; t)$ to a contour integral (clockwise):

$$e^{-\kappa x} \frac{\kappa}{2\pi} \int_{-\infty}^{\infty} \frac{e^{ik(x+8k^2t)}}{k^2 + \kappa^2} \, dk = e^{-\kappa x} \frac{\kappa}{2\pi} \oint \frac{e^{ik(x+8k^2t)}}{k^2 + \kappa^2} \, dk$$

$$= \frac{1}{2} e^{-8\kappa^3 t}$$

It shows that this integral term is independent of x and decreases in time t exponentially. Thus, this integral term is neglected to simplify the presentation. The evolution of the impulse in the KdV system is determined as

$$\phi(x, t) = 2 \frac{\partial}{\partial x} K^{(1)}(x, x; t)$$

$$= \frac{4\kappa^2 e^{-\kappa(2x - 8\kappa^2 t)}}{\left[1 + \frac{1}{2} e^{-\kappa(2x - 8\kappa^2 t)}\right]^2}$$

$$+ \frac{\kappa e^{-\kappa x} \int_{-\infty}^{\infty} \left[\frac{k+i\kappa}{k-i\kappa} + \frac{1}{2} e^{-\kappa(2x - 8\kappa^2 t)}\right] e^{ik(x+8k^2t)} \, dk}{\left[1 + \frac{1}{2} e^{-\kappa(2x - 8\kappa^2 t)}\right]^2}$$

$$= 2\kappa^2 \mathrm{sech}^2 \kappa \left(x - 4\kappa^2 t + \frac{1}{2\kappa} \ln 2\right)$$

$$\times \left\{1 + \frac{e^{-\kappa x}}{4\pi\kappa} \int_{-\infty}^{\infty} \left[\frac{k + i\kappa}{k - i\kappa} e^{\kappa(2x - 8\kappa^2 t)} + \frac{1}{2}\right] e^{ik(x+8k^2t)} \, dk\right\}$$

It indicates that an initial impulse evolves to a soliton propagating to the right and some radiation propagating to the left. Figure 4.1 illustrates the evolution.

Exercise 4.11: Show that (4.24b) is a solution of (4.22).

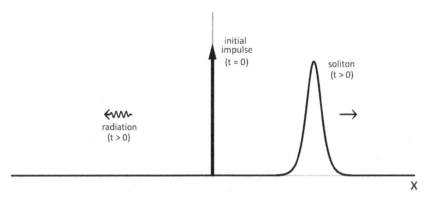

Fig. 4.1. Evolution of an impulse (at $t = 0$) to a soliton together with some radiation later.

Exercise 4.12: Explain why the radiation propagates to the left (in the $-x$ direction).

Ans: The exponential factor "$e^{ik(x+8k^2t)}$" indicates that this harmonic wave propagates in the $-x$ direction. Set $x + 8k^2t = \text{constant}$, the phase velocity of the wave is determined as $\frac{dx}{dt} = -8k^2$.

4.6.2 *Evolution of a rectangular pulse*

The initial condition is a rectangular pulse $\phi(x, 0) = A[U(x + L) - U(x - L)]$, where $U(x)$ is a unit step function and $A = (\frac{\pi}{L})^2$. The linear Schrödinger equation (4.6) becomes

$$\psi_{xx} + \lambda\psi = 0, \quad \text{for } |x| > L \tag{4.25}$$

and

$$\psi_{xx} + (\lambda + A)\psi = 0, \quad \text{for } |x| < L$$

subject to the continuity conditions at $|x| = L$:

$$\psi(|L|^+) = \psi(|L|^-) \text{ and } \psi_x(|L|^+) = \psi_x(|L|^-) \tag{4.26}$$

In the following, (4.25) is solved for the bound states (i.e., $\lambda_n = -\kappa_n^2$ and $\kappa_n^2 < A$) and for the unbound states (i.e., $\lambda = k^2$).

1. For the bound states, the eigenfunctions of (4.25) are given as

$$\psi_n = \begin{cases} \left(\frac{x}{|x|}\right)^{n-1} c_n e^{-\kappa_n |x|}, & \text{for } |x| > L \\ B_n \sin\left(\frac{n\pi}{2} - k_n x\right), & \text{for } |x| < L \end{cases} \qquad (4.27a)$$

where $k_n = \sqrt{A - \kappa_n^2}$, $n = 1, 2, \ldots, N$, N is limited by $\kappa_N^2 < A < \kappa_{N+1}^2$; the scattering amplitude c_n is determined, via the normalization $\int_{-\infty}^{\infty} \psi_n^2 \, dx = 1$. With the aid of (4.27a), the continuity conditions (4.26) set up the following equations:

$$c_n e^{-\kappa_n L} = B_n \sin\left(\frac{n\pi}{2} - k_n L\right)$$

and

$$-\kappa_n c_n e^{-\kappa_n L} = -\kappa_n B_n \cos\left(\frac{n\pi}{2} - k_n L\right)$$

which implies that

$$\sin\left(\frac{n\pi}{2} - k_n L\right) = \cos\left(\frac{n\pi}{2} - k_n L\right)$$

It is deduced as $\frac{n\pi}{2} - k_n L = \frac{\pi}{4}$, which gives $k_n = \left(n - \frac{1}{2}\right)\frac{\pi}{2L}$. Thus, $\kappa_2^2 < A = \left(\frac{\pi}{L}\right)^2 < \kappa_3^2$, indicating that there are two bound states $n = 1$ and 2.

(a) $n = 1$, having $k_1 = \frac{\pi}{4L}$ and $\kappa_1 = \frac{\sqrt{15}\pi}{4L}$, and

$$\psi_1 = \begin{cases} c_1 e^{-\kappa_1 |x|}, & \text{for } |x| > L \\ B_1 \cos k_1 x, & \text{for } |x| < L \end{cases}$$

where $B_1 = \sqrt{2} e^{-\kappa_1 L} c_1$, and c_1 is determined via the normalization

$$\int_{-\infty}^{\infty} \psi_1^2 \, dx = 2\left[\int_0^L B_1^2 \cos^2 k_1 x \, dx + \int_L^{\infty} c_1^2 e^{-2\kappa_1 x} \, dx\right]$$

$$= B_1^2 \left(1 + \frac{2}{\pi}\right) L + \frac{c_1^2}{\kappa_1} e^{-2\kappa_1 L} = 1$$

which gives

$$c_1^2(0) = \frac{e^{2\kappa_1 L}}{\left[2 + \frac{4}{\pi}\left(1 + \frac{1}{\sqrt{15}}\right)\right] L}$$

(b) $n = 2$, having $k_2 = \frac{3\pi}{4L}$ and $\kappa_2 = \frac{\sqrt{7}\pi}{4L}$, and

$$\psi_2 = \begin{cases} \left(\frac{x}{|x|}\right) c_2 e^{-\kappa_2 |x|}, & \text{for } |x| > L \\ B_2 \sin k_2 x, & \text{for } |x| < L \end{cases}$$

where $B_2 = \sqrt{2} e^{-\kappa_2 L} c_2$; the normalization of the eigenfunction

$$\int_{-\infty}^{\infty} \psi_2^2 \, dx = 2 \left[\int_0^L B_2^2 \sin^2 k_2 x \, dx + \int_L^{\infty} c_2^2 e^{-2\kappa_2 x} \, dx \right]$$

$$= B_2^2 \left(1 + \frac{2}{3\pi} \right) L + \frac{c_2^2}{\kappa_2} e^{-2\kappa_2 L} = 1$$

gives

$$c_2^2(0) = \frac{e^{2\kappa_2 L}}{\left[2 + \frac{4}{3\pi} \left(1 + \frac{3}{\sqrt{7}} \right) \right] L}$$

Exercise 4.13: Show that (4.27a) is an eigenfunction of (4.25) and meets the continuity conditions (4.26).

2. For an unbound state, a general solution of (4.25) is given as

$$\psi = \begin{cases} e^{-ikx} + b(k) e^{ikx}, & \text{for } x > L \\ a(k) e^{-ikx}, & \text{for } x < -L \\ C e^{ik_0 x} + D e^{-ik_0 x}, & \text{for } |x| < L \end{cases} \quad (4.27b)$$

where $k_0 = \sqrt{A + k^2} = \sqrt{\left(\frac{\pi}{L}\right)^2 + k^2}$.

With the aid of (4.27b), the continuity conditions (4.26) set up the following equations for the unknowns $b(k)$, $a(k)$ and C and D:

$$e^{-ikL} + b(k) e^{ikL} = C e^{ik_0 L} + D e^{-ik_0 L}$$

$$-ik[e^{-ikL} - b(k) e^{ikL}] = ik_0 [C e^{ik_0 L} - D e^{-ik_0 L}]$$

$$a(k) e^{ikL} = C e^{-ik_0 L} + D e^{ik_0 L}$$

$$-ika(k) e^{ikL} = ik_0 [C e^{-ik_0 L} - D e^{ik_0 L}]$$

These equations are solved to give

$$b(k) = 2e^{-2ikL}e^{-ik_0L}\frac{\left(1 - \frac{k_0}{k}\right) - \left(1 + \frac{k_0}{k}\right)e^{-4ik_0L}}{\left(1 - \frac{k_0}{k}\right)^2 - \left(1 + \frac{k_0}{k}\right)^2 e^{-4ik_0L}}$$

$$a(k) = -\frac{4\frac{k_0}{k}e^{-2ikL}e^{-2ik_0L}}{\left(1 - \frac{k_0}{k}\right)^2 - \left(1 + \frac{k_0}{k}\right)^2 e^{-4ik_0L}}$$

$$C = \frac{2e^{-ikL}e^{-ik_0L}\left(1 - \frac{k_0}{k}\right)}{\left(1 - \frac{k_0}{k}\right)^2 - \left(1 + \frac{k_0}{k}\right)^2 e^{-4ik_0L}}$$

$$D = -\frac{2e^{-ikL}e^{-3ik_0L}\left(1 + \frac{k_0}{k}\right)}{\left(1 - \frac{k_0}{k}\right)^2 - \left(1 + \frac{k_0}{k}\right)^2 e^{-4ik_0L}}$$

Then, (4.3) is given as

$$F(x;t) = \sum_{n=1}^{2} c_n^2(t)e^{-\kappa_n x} + \frac{1}{2\pi}\int_{-\infty}^{\infty} b(k;t)e^{ikx}\,dk$$

where $c_n^2(t) = c_n^2(0)e^{8\kappa_n^3 t}$ and $b(k;t) = b(k)e^{i8k^3 t}$. In the following, the radiation part (the integral term) of $F(x;t)$ is neglected; the GLM equation (4.6) is solved as (see (6.2))

$$K^{(2)}(x,x;t)$$

$$= -\frac{c_1^2(t)e^{-2\kappa_1 x} + c_2^2(t)e^{-2\kappa_2 x} + \frac{(\kappa_1 - \kappa_2)^2 c_1^2(t)c_2^2(t)e^{-2(\kappa_1+\kappa_2)x}}{2\kappa_1\kappa_2(\kappa_1+\kappa_2)}}{|A_{ij}^{(2)}|}$$

where

$$|A_{ij}^{(2)}| = 1 + \frac{c_1^2(t)e^{-2\kappa_1 x}}{2\kappa_1} + \frac{c_2^2(t)e^{-2\kappa_2 x}}{2\kappa_2}$$

$$+ \left(\frac{\kappa_1 - \kappa_2}{\kappa_1 + \kappa_2}\right)^2 \frac{c_1^2(t)c_2^2(t)e^{-2(\kappa_1+\kappa_2)x}}{4\kappa_1\kappa_2}$$

Then, the solution of the KdV equation (4.7) is obtained as

$$\phi(x;t) = 2\frac{\partial}{\partial x}K^{(2)}(x, x;t)$$

$$= 4\frac{\kappa_1 a_1 + \kappa_2 a_2 + \frac{(\kappa_1-\kappa_2)^2}{\kappa_1\kappa_2}a_1 a_2\left[1 + \frac{\kappa_2^3 a_1 + \kappa_1^3 a_2}{4\kappa_1\kappa_2(\kappa_1+\kappa_2)^2}\right]}{\left[1 + \frac{a_1}{2\kappa_1} + \frac{a_2}{2\kappa_2} + \left(\frac{\kappa_1-\kappa_2}{\kappa_1+\kappa_2}\right)^2 \frac{a_1 a_2}{4\kappa_1\kappa_2}\right]^2}$$

where

$$a_1(x,t) = c_1^2(t)e^{-2\kappa_1 x} = c_1^2(0)e^{-2\kappa_1(x-4\kappa_1^2 t)}$$

$$a_2(x,t) = c_2^2(t)e^{-2\kappa_2 x} = c_2^2(0)e^{-2\kappa_2(x-4\kappa_2^2 t)}$$

This solution evolves a rectangular pulse to two solitons; as $\quad t \to \infty$,

$$\phi(x;t \to \infty) = \sum_{n=1}^{2} 2\kappa_n^2 sech^2\kappa_n(x - 4\kappa_n^2 t + x_n)$$

where $x_n = \frac{1}{2\kappa_n} ln\left[\left(\frac{\kappa_1 - \kappa_2}{\kappa_1 + \kappa_2}\right)^2 \frac{2\kappa_n}{c_n^2(0)}\right]$. Two separate solitons propagate at different speeds $S_n = 4\kappa_n^2$; the amplitude is proportional to the speed so that the larger one is positioned in the front.

Exercise 4.14: Show that (4.27b) satisfies (4.25) and the continuity conditions (4.26).

Problems

P4.1. Show that (4.2a) and (4.2b) are solutions of Eq. (4.1a), where $V(x)$ is a localized potential well, and find κ_n and k.

P4.2. Set $[\psi] = \begin{bmatrix} \Theta \\ \Psi \end{bmatrix}$ to re-express (1.26) and (1.27) in the operator equation forms of (4.6c) and (4.8d), respectively, where the Lax pair $[L]$ and $[A]$ is a 2×2 matrix. Find the matrix operators $[L]$ and $[A]$ of the nonlinear Schrödinger equation (1.25a).

P4.3. With the Lax pair $[L]$ and $[A]$ defined in P4.2, show that the Lax equation (4.9a) represents the nonlinear Schrödinger equation (1.25a).

P4.4. Find the AKNS pair $[X]$ and $[T]$ so that the AKNS equation (4.12) represents the nonlinear Schrödinger equation (1.25a).

P4.5. Find the Lax pair $\vec{L}$ and $\vec{A}$ of the mKdV equation (3.15a) and show that the Lax equation (4.9) represents the mKdV equation.

P4.6. Determine $[L]$ and $[A]$, the matrix Lax pair of the mKdV equation (3.15a), and show that the matrix Lax equation (4.9a) represents the mKdV equation.

P4.7. Find the AKNS pair $[X]$ and $[T]$ so that the AKNS equation (4.12) represents the mKdV equation (3.15a).

P4.8. Show that (1.26) converts into

$$\begin{bmatrix} \Theta \\ \Psi \end{bmatrix}_{xx} + (\lambda^2 + |\phi|^2) \begin{bmatrix} \Theta \\ \Psi \end{bmatrix} = \begin{bmatrix} 0 & \phi_x \\ -\phi_x^* & 0 \end{bmatrix} \begin{bmatrix} \Theta \\ \Psi \end{bmatrix} \qquad \text{(P4.1)}$$

If ϕ_x is a real function, show that both Θ and Ψ are governed by a linear Schrödinger equation with $\Psi = \pm i\Theta$.

P4.9. With the aid of the results of P4.8 and the examples presented in Section 4.6, show the steps of applying the inverse scattering transform to analyze the evolution of a rectangular pulse in a cubic nonlinear Schrödinger system (1.25a).

Chapter 5

Basis of Inverse Scattering Transform

5.1 Derivation of the GLM Integral Equation

$$K(x,y) + F(x+y) + \int_x^\infty K(x,z)F(y+z)dz = 0 \qquad (4.5)$$

Take the inverse Fourier transform on the linear Schrödinger eigen equation of (4.1a):

$$\varphi_{xx} + [E - V(x)]\varphi = 0$$

where $V(x)$ is a localized potential function, the eigenvalue $E = \lambda^2$, and λ is the spectral parameter; and set

$$\int_{-\infty}^\infty \frac{d\lambda}{2\pi} e^{i\lambda y} \varphi(x,\lambda) = \hat{\varphi}(x,y)$$

It converts (4.1a) to (4.1b):

$$\left[\frac{\partial^2}{\partial x^2} - \frac{\partial^2}{\partial y^2} - V(x) \right] \hat{\varphi}(x,y) = 0$$

It shows that $\hat{\varphi}(x,y)$ is a solution of a linear wave equation and $V(x)$ represents a wave scatterer.

Consider first that the bound states do not exist, the solution of (4.1a) in the region to the RHS of the scatterer is given as

$$\varphi^+(x,\lambda) \sim e^{-i\lambda x} + b(\lambda)e^{i\lambda x} \quad \text{as } x \to \infty$$

where $b(\lambda)$ is the reflection coefficient.

A Jost solution for $x > 0$ is set as

$$\varphi^+(x, \lambda) = \varphi_0^+(x, \lambda) + \int_x^\infty K(x, z)\varphi_0^+(z, \lambda)dz \qquad (5.1a)$$

where

$$\varphi_0^+(x, \lambda) = e^{-i\lambda x} + b(\lambda)e^{i\lambda x}$$

Take the inverse Fourier transform on (5.1a); each term on the RHS of (5.1a) converts to

$$\int_{-\infty}^\infty \frac{d\lambda}{2\pi}e^{i\lambda y}[e^{-i\lambda x} + b(\lambda)e^{i\lambda x}] = \int_{-\infty}^\infty e^{-i\lambda(x-y)}\frac{d\lambda}{2\pi} + \hat{F}(x + y) \qquad (5.1b)$$

and

$$\int_{-\infty}^\infty \frac{d\lambda}{2\pi}e^{i\lambda y}\int_x^\infty K(x, z)[e^{-i\lambda z} + b(\lambda)e^{i\lambda z}]dz$$

$$= K(x, y) + \int_x^\infty K(x, z)\hat{F}(y + z)dz \qquad (5.1c)$$

where

$$\hat{F}(x + y) = \int_{-\infty}^\infty b(\lambda)e^{i\lambda(x+y)}\frac{d\lambda}{2\pi}$$

It leads to

$$\int_{-\infty}^\infty [\varphi(x, \lambda) - e^{-i\lambda x}]e^{i\lambda y}\frac{d\lambda}{2\pi}$$

$$= K(x, y) + \hat{F}(x + y) + \int_x^\infty K(x, z)\hat{F}(y + z)dz \qquad (5.2)$$

The integral on the left-hand side (LHS) of (5.2) is converted to a contour integration. The contour of the integration is in the upper half of the complex λ-plane comprised of a semicircular arc of radius R and the real line segment $(-R, R)$ for $R \to \infty$. Since $\text{Im}(\lambda) > 0$ and $y > x$, $e^{i\lambda(y-x)}$ on the arc decays exponentially with R; the added integration along the arc approaches zero as $R \to \infty$. The contour integration is zero because $\varphi(x, \lambda)e^{i\lambda x} - 1$ has no poles, that is,

$$\int_{-\infty}^{\infty} [\varphi(x, \lambda) - e^{-i\lambda x}]e^{i\lambda y}\frac{d\lambda}{2\pi} = \oint [\varphi(x, \lambda)e^{i\lambda x} - 1]e^{i\lambda(y-x)}\frac{d\lambda}{2\pi}$$

$$= 0 \quad \text{for } y > x$$

The Gelfand–Levitan–Marchenko (GLM) linear integral equation is derived as

$$K(x, y) + \hat{F}(x + y) + \int_{x}^{\infty} K(x, z)\hat{F}(y + z)dz = 0 \quad \text{for } y > x$$

$$(5.3)$$

In the existence of bound states, i.e., $\varphi(x, \lambda)e^{i\lambda x} - 1$ contains discrete poles in the upper half λ-plane, the contour integral

$$\oint [\varphi(x, \lambda)e^{i\lambda x} - 1]e^{i\lambda(y-x)}\frac{d\lambda}{2\pi} \neq 0$$

The eigenfunction of (4.1a), for a bound state with eigenenergy $E_n = -\kappa_n^2$, in the region to the RHS of the scatterer is given as

$$\varphi_n^+(x, \lambda = i\kappa_n) = \bar{\varphi}_n^+ \sim c_n e^{-\kappa_n x} = \varphi_{n0}^+(x, i\kappa_n) \quad \text{as } x \to \infty$$

where the overbar of φ stands for $\lambda = i\kappa_n$ and c_n is a scattering amplitude.

A Jost solution for a bound state φ_n^+ is set as

$$\bar{\varphi}_n^+ = \varphi_n^+(x, \lambda = i\kappa_n) = \varphi_{n0}^+(x, i\kappa_n) + \int_{x}^{\infty} K(x, z)\varphi_{n0}^+(z, i\kappa_n)dz$$

$$= c_n \left[e^{-\kappa_n x} + \int_{x}^{\infty} K(x, z)e^{-\kappa_n z}dz \right] \quad (5.4)$$

Then,

$$\oint [\varphi(x, \lambda)e^{i\lambda x} - 1]e^{i\lambda(y-x)}\frac{d\lambda}{2\pi}$$

$$= \oint \varphi(x, \lambda)e^{i\lambda y}\frac{d\lambda}{2\pi} = i\sum_{n=1}^{N} R_n = i\sum_{n=1}^{N} ic_n\bar{\varphi}_n^+ e^{-\kappa_n y}$$

$$= -\sum_{n=1}^{N} c_n^2 e^{-\kappa_n y} \left[e^{-\kappa_n x} + \int_{x}^{\infty} K(x, z)e^{-\kappa_n z}dz \right] \quad (5.5)$$

where R_n is the residue of $\varphi(x, \lambda)e^{i\lambda y}$ at $\lambda = i\kappa_n$. Derivation of R_n is presented in Section 5.2.

Thus, (5.2) becomes

$$K(x, y) + \int_{-\infty}^{\infty} b(\lambda)e^{i\lambda(x+y)} \frac{d\lambda}{2\pi}$$

$$+ \int_{x}^{\infty} \left[K(x, z) \int_{-\infty}^{\infty} b(\lambda)e^{i\lambda(y+z)} \frac{d\lambda}{2\pi} \right] dz$$

$$= -\sum_{n=1}^{N} c_n^2 e^{-\kappa_n y} \left[e^{-\kappa_n x} + \int_{x}^{\infty} K(x, z)e^{-\kappa_n z} dz \right] \qquad (5.6a)$$

Set

$$F(x + y) = \int_{-\infty}^{\infty} b(\lambda)e^{i\lambda(x+y)} \frac{d\lambda}{2\pi} + \sum_{n=1}^{N} c_n^2 e^{-\kappa_n(x+y)}$$

Equation (5.6a) is re-expressed as

$$K(x, y) + F(x + y) + \int_{x}^{\infty} K(x, z)F(y + z)dz = 0, \quad \text{for } y > x$$

$$(5.6b)$$

Equation (5.6b) has the same form as (5.3), except the scattering function $F(x + y)$ also includes the contribution from the discrete scattering spectrum (bound states). This is the GLM integral equation, which applies the scattering data S to reconstruct the scatterer V. The scattering data, comprised of the reflection coefficient, $b(\lambda)$, the discrete eigenvalues, $-\kappa_n^2$, and the corresponding scattering amplitudes, c_n, are used to define $F(x + y)$, (5.6b) is then solved; the solution $K(x, y)$ determines $V(x)$ through the relation $V(x) = -2K_x(x, x)$.

Exercise 5.1: Show that (5.6a) is re-expressed as (5.6b).

5.2   Derivation of the Residues of $\varphi(x, \lambda)e^{i\lambda y}$

Introduce Jost solutions $\psi_+(x, \lambda)$ and $\psi_-(x, \lambda)$ of (4.1a) as follows:

$$\psi_+(x, \lambda) = e^{i\lambda x} + \int_{x}^{\infty} K(x, z)e^{i\lambda z} dz \qquad (5.7a)$$

and

$$\psi_-(x,\lambda) = e^{-i\lambda x} + \int_{-\infty}^{x} L(x,z)e^{-i\lambda z}dz \qquad (5.7b)$$

where $\psi_+(x,\lambda) \sim e^{i\lambda x}$ as $x \to \infty$ and $\psi_-(x,\lambda) \sim e^{-i\lambda x}$ as $x \to -\infty$. Thus, $\varphi(x,\lambda)$ can be written as

$$\varphi(x,\lambda) = \psi_+^*(x,\lambda) + b(\lambda)\psi_+(x,\lambda) = a(\lambda)\psi_-(x,\lambda) \qquad (5.8)$$

It leads to

$$\psi_- = a^{-1}\psi_+^* + ba^{-1}\psi_+ \qquad (5.9)$$

At a pole $\lambda = i\kappa_n$ of $a(\lambda)$, the associated discrete eigenfunction $\psi_n(x,i\kappa_n)$ can be represented as

$$\psi_n(x,i\kappa_n) = c_n\psi_+(x,i\kappa_n) = d_n\psi_-(x,i\kappa_n) \text{ i.e., } c_n\bar{\psi}_+ = d_n\bar{\psi}_- \tag{5.10}$$

where the overbar of ψ_+ and ψ_- stands for $\lambda = i\kappa_n$, c_n is a scattering amplitude, and d_n is an unknown constant.

From (4.1a),

$$\psi_-\psi_{+xx} + [E - V(x)]\psi_+\psi_- = 0$$

and

$$\psi_+\psi_{-xx} + [E - V(x)]\psi_+\psi_- = 0$$

Then,

$$\psi_-\psi_{+xx} - \psi_+\psi_{-xx} = 0 = (\psi_-\psi_{+x} - \psi_+\psi_{-x})_x$$

and thus,

$$W(\psi_-,\psi_+) = \psi_-\psi_{+x} - \psi_+\psi_{-x} = \text{constant}$$

$$= (\psi_-\psi_{+x} - \psi_+\psi_{-x})|_{x=\infty} = 2i\lambda a^{-1} \qquad (5.11)$$

where "W" stands for Wronskian. Taking a λ derivative on both sides of (5.11) yields

$$\frac{\partial}{\partial\lambda}W(\psi_-,\psi_+) = W(\psi_{-\lambda},\psi_+) + W(\psi_-,\psi_{+\lambda})$$

$$= 2i\left(a^{-1} - \lambda\frac{a_\lambda}{a^2}\right) \qquad (5.12)$$

Since the RHS of (5.12) is independent of x, the Wronskians on the LHS of (5.12) can be evaluated at $x = -\infty$. With the aid of the relation (5.10), (5.12) with $\lambda = i\kappa_n$ becomes

$$W\left(\bar{\psi}_{-\lambda}, \frac{d_n}{c_n}\bar{\psi}_{-}\right)\Bigg|_{x=-\infty} + W\left(\frac{c_n}{d_n}\bar{\psi}_{+}, \bar{\psi}_{+\lambda}\right)\Bigg|_{x=-\infty}$$

$$= 2\kappa_n\left(\frac{a_\lambda}{a^2}\right)\Bigg|_{\lambda=i\kappa_n} = -\frac{c_n}{d_n}W(\bar{\psi}_{+\lambda}, \bar{\psi}_{+})\Bigg|_{x=-\infty} \qquad (5.13)$$

where $(a^{-1})|_{\lambda=i\kappa_n} = 0$ because $i\kappa_n$ is a pole of $a(\lambda)$; the relations $W(\phi, \psi) = -W(\psi, \phi)$ and $W(\alpha\phi, \beta\psi) = \alpha\beta W(\phi, \psi)$ are applied; and $W(\bar{\psi}_{-\lambda}, \bar{\psi}_{-})|_{x=-\infty} = 0$ as evaluated directly through (5.7b).

Again, with the aid of (4.1a), it obtains

$$\varphi_\lambda\varphi_{xx} - \varphi\varphi_{xx\lambda} = 2\lambda\varphi^2 = (\varphi_\lambda\varphi_x - \varphi\varphi_{x\lambda})_x$$

Set $\varphi = \psi_+(x, i\kappa_n) = \bar{\psi}_+$ and take integration on x from $-\infty$ to ∞, it yields

$$(\bar{\psi}_{+\lambda}\bar{\psi}_{+x} - \bar{\psi}_{+}\bar{\psi}_{+x\lambda})|_{x=\infty} - (\bar{\psi}_{+\lambda}\bar{\psi}_{+x} - \bar{\psi}_{+}\bar{\psi}_{+x\lambda})|_{x=-\infty}$$

$$= W(\bar{\psi}_{+\lambda}, \bar{\psi}_{+})|_{x=\infty} - W(\bar{\psi}_{+\lambda}, \bar{\psi}_{+})|_{x=-\infty} = \frac{2i\kappa_n}{c_n^2}$$

$$= -W(\bar{\psi}_{+\lambda}, \bar{\psi}_{+})|_{x=-\infty} \qquad (5.14)$$

where $W(\bar{\psi}_{+\lambda}, \bar{\psi}_{+})|_{x=\infty} = 0$ as evaluated directly through (5.7a).

The relations (5.13) and (5.14) give

$$\left(\frac{a_\lambda}{a^2}\right)\Bigg|_{\lambda=i\kappa_n} = \frac{i}{c_n d_n} \qquad (5.15)$$

The residues of $a(\lambda)$ are obtained as

$$\oint a(\lambda)d\lambda = 2\pi i \times \{\text{sum of the residues of } a(\lambda)\}$$

$$= -2\pi i \sum_{n=1}^{N}\left(\frac{a^2}{a_\lambda}\right)\Bigg|_{\lambda=i\kappa_n} = -\sum_{n=1}^{N} 2\pi c_n d_n$$

The contour integral is then evaluated as

$$\oint \varphi(x,\lambda)e^{i\lambda y}\frac{d\lambda}{2\pi} = \oint a(\lambda)\psi_-(x,\lambda)e^{i\lambda y}\frac{d\lambda}{2\pi}$$

$$= -\sum_{n=1}^{N} c_n d_n \bar{\psi}_- e^{-\kappa_n y} = -\sum_{n=1}^{N} c_n^2 \bar{\psi}_+ e^{-\kappa_n y}$$

$$= -\sum_{n=1}^{N} c_n^2 e^{-\kappa_n y} \left[e^{-\kappa_n x} + \int_x^{\infty} K(x,z)e^{-\kappa_n z}dz \right]$$

Exercise 5.2: Show that $W(\bar{\psi}_{+\lambda}, \bar{\psi}_+)|_{x=\infty} = 0 = W(\bar{\psi}_{-\lambda}, \bar{\psi}_-)|_{x=-\infty}$.

Ans: As $\to x \pm \infty$, (5.7a) and (5.7b) reduce to $\psi_\pm(x,\lambda) = e^{\pm i\lambda x}$.

5.3 Proof of the Jost Solutions Satisfying the Linear Schrödinger Eigen Equation

The Jost solution (5.1a) is expressed in terms of the kernel $K(x,z)$ of the GLM linear integral equation; a proof of (5.1a) being a solution of the linear Schrödinger equation, it is proceeded by first showing that $K(x,y)$ satisfies (4.1b). Thus, the differential operator $\left[\frac{\partial^2}{\partial x^2} - \frac{\partial^2}{\partial y^2} - V(x)\right]$ of (4.1b) is applied to the GLM equation (4.5).

With the aid of

$$\frac{\partial^2}{\partial x^2} \int_x^{\infty} K(x,z)F(y+z)dz = \int_x^{\infty} K_{xx}(x,z)F(y+z)dz$$

$$- [2K_x(x,x) - K_z(x,z)|_{z=x}]F(x+y) - K(x,x)F_x(x+y)$$

$$\left[\frac{\partial^2}{\partial x^2} - \frac{\partial^2}{\partial y^2} - V(x)\right]F(x+y) = -V(x)F(x+y)$$

where the relation $F_{xx}(x+y) = F_{yy}(x+y)$ is applied; and

$$\frac{\partial^2}{\partial y^2} \int_x^{\infty} K(x,z)F(y+z)dz$$

$$= \int_x^\infty K(x,z) F_{zz}(y+z) dz = \int_x^\infty K_{zz}(x,z) F(y+z) dz$$

$$- K(x,x) F_x(x+y) + K_z(x,z)|_{z=x} F(x+y)$$

where integration by parts is applied twice on the integral of $K(x,z) F_{zz}(y+z)$.

It leads to

$$\left[\frac{\partial^2}{\partial x^2} - \frac{\partial^2}{\partial y^2} - V(x) \right] \left[K(x,y) + F(x+y) \right.$$

$$\left. + \int_x^\infty K(x,z) F(y+z) dz \right] = \left[\frac{\partial^2}{\partial x^2} - \frac{\partial^2}{\partial y^2} - V(x) \right] K(x,y)$$

$$+ \int_x^\infty \left\{ \left[\frac{\partial^2}{\partial x^2} - \frac{\partial^2}{\partial z^2} - V(x) \right] K(x,z) \right\} F(y+z) dz$$

$$- [V(x) + 2K_x(x,x)] F(x+y)$$

It shows that if

$$K(x,y) + F(x+y) + \int_x^\infty K(x,z) F(y+z) dz = 0 \quad \text{for } y \geq x$$

and

$$V(x) = -2K_x(x,x)$$

Then,

$$\left[\frac{\partial^2}{\partial x^2} - \frac{\partial^2}{\partial y^2} - V(x) \right] K(x,y) = 0 \qquad (5.16)$$

which is the same equation as (4.1b). In other words, if $K(x,y)$ is a solution of the GLM equation and defines the potential function $V(x)$ of the linear Schrödinger equation (4.1a), then it is also a solution of (4.1b).

Next, consider first that the bound states do not exist, the Jost solution of (4.1a) in the region to the RHS of the scatterer is given as

$$\varphi^+(x,\lambda) = e^{-i\lambda x} + b(\lambda) e^{i\lambda x} + \int_x^\infty K(x,z)[e^{-i\lambda z} + b(\lambda) e^{i\lambda z}] dz$$

$$(5.17)$$

The second derivative on the integral term on the RHS of the Jost solution (5.17) gives

$$
\frac{\partial^2}{\partial x^2} \int_x^\infty K(x,z)[e^{-i\lambda z} + b(\lambda)e^{i\lambda z}]dz
$$

$$
= \int_x^\infty K_{xx}(x,z)[e^{-i\lambda z} + b(\lambda)e^{i\lambda z}]dz - 2K_x(x,x)[e^{-i\lambda x} + b(\lambda)e^{i\lambda x}]
$$

$$
+ i\lambda \left[K(x,x) - \frac{i}{\lambda}K_z(x,z) \Big|_{z=x} \right] e^{-i\lambda x}
$$

$$
- i\lambda \left[K(x,x) + \frac{i}{\lambda}K_z(x,z) \Big|_{z=x} \right] b(\lambda)e^{i\lambda x} \tag{5.18}
$$

With the aid of the relations $[e^{-i\lambda z} + b(\lambda)e^{i\lambda z}] = \frac{i}{\lambda}[e^{-i\lambda z} - b(\lambda)e^{i\lambda z}]_z = -\frac{1}{\lambda^2}[e^{-i\lambda z} + b(\lambda)e^{i\lambda z}]_{zz}$, integration by parts is applied twice on the RHS integral term of (5.17); this integral term becomes

$$
\int_x^\infty K(x,z)[e^{-i\lambda z} + b(\lambda)e^{i\lambda z}]dz
$$

$$
= -\frac{1}{\lambda^2} \int_x^\infty K_{zz}(x,z)[e^{-i\lambda z} + b(\lambda)e^{i\lambda z}]dz
$$

$$
- \frac{i}{\lambda} \left[K(x,x) - \frac{i}{\lambda}K_z(x,z) \Big|_{z=x} \right] e^{-i\lambda x}
$$

$$
+ \frac{i}{\lambda} \left[K(x,x) + \frac{i}{\lambda}K_z(x,z) \Big|_{z=x} \right] b(\lambda)e^{i\lambda x} \tag{5.19}
$$

Apply the differential operator $(\frac{\partial^2}{\partial x^2} + E)$ to (5.17); with the aid of (5.18) and (5.19), it leads to

$$
\left(\frac{\partial^2}{\partial x^2} + E \right) \varphi^+(x,\lambda)
$$

$$
= \frac{\partial^2}{\partial x^2} \int_x^\infty K(x,z)[e^{-i\lambda z} + b(\lambda)e^{i\lambda z}]dz
$$

$$
+ \lambda^2 \int_x^\infty K(x,z)[e^{-i\lambda z} + b(\lambda)e^{i\lambda z}]dz
$$

$$= \int_x^\infty [K_{xx}(x, z) - K_{zz}(x, z)][e^{-i\lambda z} + b(\lambda)e^{i\lambda z}]dz$$

$$- 2K_x(x, x)[e^{-i\lambda x} + b(\lambda)e^{i\lambda x}] \tag{5.20}$$

where $E = \lambda^2$ is applied.

Then,

$$\left\{ \frac{\partial^2}{\partial x^2} + [E - V(x)] \right\} \varphi^+(x, \lambda)$$

$$= \int_x^\infty [K_{xx}(x, z) - K_{zz}(x, z) - V(x)K(x, z)][e^{-i\lambda z} + b(\lambda)e^{i\lambda z}]dz$$

$$- [V(x) + 2K_x(x, x)][e^{-i\lambda x} + b(\lambda)e^{i\lambda x}] \tag{5.21}$$

It shows that

$$\left\{ \frac{\partial^2}{\partial x^2} + [E - V(x)] \right\} \varphi^+(x, \lambda) = 0$$

if $V(x) = -2K_x(x, x)$ and $K(x, y)$ satisfies (5.16), i.e., $K(x, y)$ is the solution of the GLM integral equation. A proof for (5.4) can be done in the same way.

Exercise 5.3: Prove that (5.4) is a Jost solution of (4.1a) for a bound state.

Ans: Substitute $c_n e^{-\kappa_n x}$ for $[e^{-i\lambda z} + b(\lambda)e^{i\lambda z}]$ and $-\kappa_n^2$ for λ^2 in (5.20).

Chapter 6

Solitary Waves

6.1 Illustration of IST via Solving the KdV Equation

IST is applied to the KdV equation (4.7) to look for soliton solutions; initial conditions, which grant reflectionless potential in (4.6), are imposed. Thus, $b(k; 0) = 0$ and (4.3) reduces to

$$F(x; t) = \sum_{n=1}^{N} c_n^2(0) e^{8\kappa_n^3 t - \kappa_n x} \tag{4.18a}$$

Based on the formulations presented in Chapter 4, solutions of the KdV equation up to four solitons are elicited for exemplifying the IST applying procedure. First, the results of (4.21) are presented for $N = 1, 2, 3$ and 4:

1. $N = 1; \kappa_1 = \kappa$

$$H_1(x; t) = -\frac{2\kappa c_1^2(t) e^{-\kappa x}}{2\kappa + c_1^2(t) e^{-2\kappa x}} = -\frac{2\kappa c_1^2(0) e^{-(\kappa x - 8\kappa^3 t)}}{2\kappa + c_1^2(0) e^{-2(\kappa x - 4\kappa^3 t)}}$$

then,

$$K^{(1)}(x, x; t) = -\frac{2\kappa c_1^2(0) e^{-2\kappa(x - 4\kappa^2 t)}}{2\kappa + c_1^2(0) e^{-2\kappa(x - 4\kappa^2 t)}} \tag{6.1}$$

2. $N = 2$

$$H_1(x;t) = \frac{A_{22}C_1 - A_{12}C_2}{A_{11}A_{22} - A_{12}A_{21}}$$

$$= \frac{-c_1^2(t)\left(1 + \frac{(\kappa_1 - \kappa_2)c_2^2(t)e^{-2\kappa_2 x}}{2\kappa_2(\kappa_1 + \kappa_2)}\right)e^{-\kappa_1 x}}{\left|A_{ij}^{(2)}\right|}$$

$$H_2(x;t) = \frac{A_{11}C_2 - A_{21}C_1}{A_{11}A_{22} - A_{12}A_{21}}$$

$$= \frac{-c_2^2(t)\left(1 - \frac{(\kappa_1 - \kappa_2)c_1^2(t)e^{-2\kappa_1 x}}{2\kappa_1(\kappa_1 + \kappa_2)}\right)e^{-\kappa_2 x}}{\left|A_{ij}^{(2)}\right|}$$

where

$$A_{11} = 1 + \frac{c_1^2(t)e^{-2\kappa_1 x}}{2\kappa_1}, \quad A_{12} = \frac{c_1^2(t)e^{-(\kappa_1 + \kappa_2)x}}{\kappa_1 + \kappa_2}$$

$$A_{21} = \frac{c_2^2(t)e^{-(\kappa_1 + \kappa_2)x}}{\kappa_1 + \kappa_2}, \quad A_{22} = 1 + \frac{c_2^2(t)e^{-2\kappa_2 x}}{2\kappa_2}$$

$$C_1 = -c_1^2(t)e^{-\kappa_1 x} \quad \text{and} \quad C_2 = -c_2^2(t)e^{-\kappa_2 x}, \quad \text{and}$$

$$\left|A_{ij}^{(2)}\right| = 1 + \frac{c_1^2(t)e^{-2\kappa_1 x}}{2\kappa_1} + \frac{c_2^2(t)e^{-2\kappa_2 x}}{2\kappa_2}$$

$$+ \left(\frac{\kappa_1 - \kappa_2}{\kappa_1 + \kappa_2}\right)^2 \frac{c_1^2(t)c_2^2(t)e^{-2(\kappa_1 + \kappa_2)x}}{4\kappa_1\kappa_2}$$

Then,

$$K^{(2)}(x, x; t)$$

$$= -\frac{c_1^2(t)e^{-2\kappa_1 x} + c_2^2(t)e^{-2\kappa_2 x} + \frac{(\kappa_1 - \kappa_2)^2 c_1^2(t)c_2^2(t)e^{-2(\kappa_1 + \kappa_2)x}}{2\kappa_1\kappa_2(\kappa_1 + \kappa_2)}}{\left|A_{ij}^{(2)}\right|}$$

$$= \frac{\partial}{\partial x}\ln\left|A_{ij}^{(2)}\right| \tag{6.2}$$

3. $N = 3$

$$H_1(x;t) = \frac{B_{11}C_1 + B_{12}C_2 + B_{13}C_3}{\left|A_{ij}^{(3)}\right|}$$

$$H_2(x;t) = \frac{B_{21}C_1 + B_{22}C_2 + B_{23}C_3}{\left|A_{ij}^{(3)}\right|}$$

$$H_3(x;t) = \frac{B_{31}C_1 + B_{32}C_2 + B_{33}C_3}{\left|A_{ij}^{(3)}\right|}$$

where

$$B_{11} = A_{22}A_{33} - A_{23}A_{32}$$
$$= 1 + \frac{c_2^2(t)e^{-2\kappa_2 x}}{2\kappa_2} + \frac{c_3^2(t)e^{-2\kappa_3 x}}{2\kappa_3} + \frac{(\kappa_2 - \kappa_3)^2}{(\kappa_2 + \kappa_3)^2}$$
$$\times \frac{c_2^2(t)c_3^2(t)e^{-2(\kappa_2 + \kappa_3)x}}{4\kappa_2\kappa_3}$$

$$B_{12} = A_{13}A_{32} - A_{12}A_{33}$$
$$= -\frac{c_1^2(t)e^{-(\kappa_1 + \kappa_2)x}}{\kappa_1 + \kappa_2}$$
$$\times \left(1 + \frac{\left[\kappa_3^2 + \kappa_1\kappa_2 - \kappa_3(\kappa_1 + \kappa_2)\right]c_3^2(t)e^{-2\kappa_3 x}}{2\kappa_3(\kappa_1 + \kappa_3)(\kappa_2 + \kappa_3)}\right)$$

$$B_{13} = A_{12}A_{23} - A_{13}A_{22}$$
$$= -\frac{c_1^2(t)e^{-(\kappa_1 + \kappa_3)x}}{\kappa_1 + \kappa_3}$$
$$\times \left(1 + \frac{\left[\kappa_2^2 + \kappa_1\kappa_3 - \kappa_2(\kappa_1 + \kappa_3)\right]c_2^2(t)e^{-2\kappa_2 x}}{2\kappa_2(\kappa_1 + \kappa_2)(\kappa_2 + \kappa_3)}\right)$$

$$B_{21} = A_{23}A_{31} - A_{21}A_{33}$$
$$= -\frac{c_2^2(t)e^{-(\kappa_1 + \kappa_2)x}}{\kappa_1 + \kappa_2}$$
$$\times \left(1 + \frac{\left[\kappa_3^2 + \kappa_1\kappa_2 - \kappa_3(\kappa_1 + \kappa_2)\right]c_3^2(t)e^{-2\kappa_3 x}}{2\kappa_3(\kappa_1 + \kappa_3)(\kappa_2 + \kappa_3)}\right)$$

$$B_{22} = A_{11}A_{33} - A_{13}A_{31}$$

$$= 1 + \frac{c_1^2(t)e^{-2\kappa_1 x}}{2\kappa_1} + \frac{c_3^2(t)e^{-2\kappa_3 x}}{2\kappa_3} + \frac{(\kappa_1 - \kappa_3)^2}{(\kappa_1 + \kappa_3)^2}$$

$$\times \frac{c_1^2(t)c_3^2(t)e^{-2(\kappa_1+\kappa_3)x}}{4\kappa_1\kappa_3}$$

$$B_{23} = A_{13}A_{21} - A_{11}A_{23}$$

$$= -\frac{c_2^2(t)e^{-(\kappa_2+\kappa_3)x}}{\kappa_2 + \kappa_3}$$

$$\times \left(1 + \frac{\left[\kappa_1^2 + \kappa_2\kappa_3 - \kappa_1(\kappa_2 + \kappa_3) \right] c_1^2(t)e^{-2\kappa_1 x}}{2\kappa_1(\kappa_1 + \kappa_2)(\kappa_1 + \kappa_3)} \right)$$

$$B_{31} = A_{21}A_{32} - A_{22}A_{31}$$

$$= -\frac{c_3^2(t)e^{-(\kappa_1+\kappa_3)x}}{\kappa_1 + \kappa_3}$$

$$\times \left(1 + \frac{\left[\kappa_2^2 + \kappa_1\kappa_3 - \kappa_2(\kappa_1 + \kappa_3) \right] c_2^2(t)e^{-2\kappa_2 x}}{2\kappa_2(\kappa_1 + \kappa_2)(\kappa_2 + \kappa_3)} \right)$$

$$B_{32} = A_{12}A_{31} - A_{11}A_{32}$$

$$= -\frac{c_3^2(t)e^{-(\kappa_2+\kappa_3)x}}{\kappa_2 + \kappa_3}$$

$$\times \left(1 + \frac{\left[\kappa_1^2 + \kappa_2\kappa_3 - \kappa_1(\kappa_2 + \kappa_3) \right] c_1^2(t)e^{-2\kappa_1 x}}{2\kappa_1(\kappa_1 + \kappa_2)(\kappa_1 + \kappa_3)} \right)$$

$$B_{33} = A_{11}A_{22} - A_{12}A_{21}$$

$$= 1 + \frac{c_1^2(t)e^{-2\kappa_1 x}}{2\kappa_1} + \frac{c_2^2(t)e^{-2\kappa_2 x}}{2\kappa_2}$$

$$+ \frac{(\kappa_1 - \kappa_2)^2}{(\kappa_1 + \kappa_2)^2} \frac{c_1^2(t)c_2^2(t)e^{-2(\kappa_1+\kappa_2)x}}{4\kappa_1\kappa_2}$$

Thus,

$$B_{11}C_1 + B_{12}C_2 + B_{13}C_3$$

$$= -\left(1 - \frac{c_2^2(t)e^{-4x}}{12} - \frac{c_3^2(t)e^{-6x}}{12} \right) c_1^2(t)e^{-x}$$

$$
-\frac{c_1^2(t)c_2^2(t)c_3^2(t)}{3600}e^{-11x}
$$

$$
B_{21}C_1 + B_{22}C_2 + B_{23}C_3
$$

$$
= -\left(1 + \frac{c_1^2(t)e^{-2x}}{6} - \frac{c_3^2(t)e^{-6x}}{30}\right)c_2^2(t)e^{-2x}
$$

$$
+ \frac{c_1^2(t)c_2^2(t)c_3^2(t)e^{-10x}}{720}
$$

$$
B_{31}C_1 + B_{32}C_2 + B_{33}C_3
$$

$$
= -\left(1 + \frac{c_1^2(t)e^{-2x}}{4} + \frac{c_2^2(t)e^{-4x}}{20}\right)c_3^2(t)e^{-3x}
$$

$$
- \frac{c_1^2(t)c_2^2(t)c_3^2(t)e^{-9x}}{720}
$$

$$
\left|A_{ij}^{(3)}\right| = A_{11}B_{11} + A_{12}B_{21} + A_{13}B_{31}
$$

$$
= 1 + \frac{c_1^2(t)e^{-2x}}{2} + \frac{c_2^2(t)e^{-4x}}{4} + \frac{c_3^2(t)e^{-6x}}{6}
$$

$$
+ \frac{c_1^2(t)c_2^2(t)e^{-6x}}{72} + \frac{c_1^2(t)c_3^2(t)e^{-8x}}{48}
$$

$$
+ \frac{c_2^2(t)c_3^2(t)e^{-10x}}{600} + \frac{c_1^2(t)c_2^2(t)c_3^2(t)e^{-12x}}{43200}
$$

where $\kappa_n = n, n = 1, 2$, and 3, is set in advance to simplify the expressions. Then,

$$
K^{(3)}(x, x; t) = \sum_{n=1}^{3} H_n(x; t)e^{-\kappa_n x} = \frac{\partial}{\partial x}\ln\left|A_{ij}^{(3)}\right|
$$

$$
= -\left[\left(c_1^2(t)e^{-2x} + c_2^2(t)e^{-4x} + c_3^2(t)e^{-6x}\right.\right.
$$

$$
+ \frac{c_1^2(t)c_2^2(t)e^{-6x}}{12} + \frac{c_1^2(t)c_3^2(t)e^{-8x}}{6} + \frac{c_2^2(t)c_3^2(t)e^{-10x}}{60}\left.\right)
$$

$$
\left.+ \frac{c_1^2(t)c_2^2(t)c_3^2(t)e^{-12x}}{3600}\right]\Bigg/\left|A_{ij}^{(3)}\right| \tag{6.3}
$$

4. $N = 4$

$$K^{(4)}(x, x; t) = \sum_{n=1}^{4} H_n(x; t) e^{-\kappa_n x} = \frac{\partial}{\partial x} \ln |A_{ij}^{(4)}|$$

6.2   Steps of Applying IST

So far, the GLM equation (4.5) is solved analytically, and the second auxiliary equation is applied to determine the time dependencies of the scattering data of the KdV equation. The remaining steps for solving (4.7) via IST, for four cases of $N = 1$ to 4, are illuminated in the following:

Step 1: Solve (4.6) to find the (initial) eigenfunctions $\psi_n(x)$, i.e., $\phi = \phi(x, 0)$, and the corresponding eigenvalues λ_n which are time constants.

Equation (4.6) is re-expressed through a transform $s = \tanh x$, which maps x from $(-\infty, \infty)$ to s from $(-1, 1)$ With the aid of

$$\frac{d^2}{dx^2} \to (1 - s^2)^2 \frac{d^2}{ds^2} - 2s(1 - s^2)\frac{d}{ds} = (1 - s^2)\frac{d}{ds}(1 - s^2)\frac{d}{ds}$$

and set $U(s, 0) = C\psi(x, 0)$ (4.6) is transformed to

$$\frac{d}{ds}(1 - s^2)\frac{d}{ds}U + \frac{\phi + \lambda}{(1 - s^2)}U = 0 \tag{6.4a}$$

Consider the cases with the initial conditions $\phi(x, 0) = N(N+1)\operatorname{sech}^2(x) = N(N+1)(1 - s^2)$; then, (6.4a) becomes an associated Legendre differential equation, which has known analytical solutions:

$$\frac{d}{ds}(1 - s^2)\frac{d}{ds}U_n + \left[N(N+1) - \frac{n^2}{(1 - s^2)}\right]U_n = 0 \tag{6.4b}$$

where $n^2 = -\lambda_n = \kappa_n^2$ and $0 < n \leq N$. Thus, $n = 1, 2, \ldots, N$ The normalized initial eigenfunction $\psi_n(x, 0)$ is proportional to the associated Legendre polynomial $P_N^n = U_n(s, 0)$ which is

$$P_N^n(s) = (-1)^n (1 - s^2)^{\frac{n}{2}} \frac{d^n}{ds^n} P_N(s)$$

$$= \frac{(-1)^n}{2^N N!} (1 - s^2)^{\frac{n}{2}} \frac{d^{N+n}}{ds^{N+n}} (s^2 - 1)^N \qquad (6.4c)$$

where

$$P_N(s) = \frac{1}{2^N N!} \frac{d^N}{ds^N} (s^2 - 1)^N$$

As $x \to \pm\infty$, $s \to \pm 1$; because $P_N^n(\pm 1) = 0$, which corresponds to (4.2a) and excludes (4.2b). In other words, there are only bound states existing in this class of initial conditions, which represent reflectionless potentials of the Schrödinger equation (4.6) and are characterized by $b(k) = 0$.

The normalized eigenfunction

$$\varphi_n(x, \kappa_n) = \psi_n(x) = \frac{1}{C} P_N^n(s = \tanh x) \qquad (6.4d)$$

where

$$C = \pm \left(\int_{-\infty}^{\infty} [P_N^n(\tanh x)]^2 \, dx \right)^{1/2} = \pm \left(\int_{-1}^{1} \frac{[P_N^n(s)]^2}{1 - s^2} \, ds \right)^{\frac{1}{2}} \qquad (6.4e)$$

Exercise 6.1: Show that the linear Schrödinger equation (4.6) is transformed to (6.4a).

Step 2: Apply the asymptotic relation (4.2a) to the normalized initial eigenfunction (6.4d) to determine the initial scattering data $c_n(0)$, which completes the scattering function $F(x; t)$ of (4.18a) to realize the solution $K(x, x; t)$ of the GLM equation.

1. $N = 1$ (one-soliton solution), having the initial condition $\phi^{(1)}(x, 0) = 2 \operatorname{sech}^2 x$.

For $N = 1$, $n = 1$, Eq. (6.4a) has an eigenvalue $\lambda_1 = -1$, giving $\kappa_1 = 1$. The eigenfunction of (6.4b) is the associated Legendre polynomial P_1^1 which is

$$P_1^1 = -\frac{1}{2}(1 - s^2)^{\frac{1}{2}}\frac{d^2}{ds^2}(s^2 - 1) = -(1 - s^2)^{\frac{1}{2}} = -\mathrm{sech}x$$

substitute it into (6.4e), $C = \sqrt{2}$. Thus, normalized eigenfunctions are obtained as

$$\varphi_1(x, \kappa_1) = \mp\frac{1}{\sqrt{2}}\,\mathrm{sech}x$$

Apply the asymptotic relation (4.2a), it yields

$$c_1(0) = \lim_{x\to\infty}\varphi_1(x, \kappa_1)e^{\kappa_1 x} = \lim_{x\to\infty}\left[\mp\frac{1}{\sqrt{2}}\,\mathrm{sech}x e^x\right] = \mp\sqrt{2}$$

The kernel function (6.1), for $N = 1$, is then determined explicitly as

$$K^{(1)}(x, x; t) = -2\frac{e^{8t-2x}}{(1 + e^{8t-2x})} \tag{6.1a}$$

2. $N = 2$ (two-solitons solution), having the initial condition $\phi^{(2)}(x, 0) = 6\,\mathrm{sech}^2 x$.

In the case of $N = 2$, $n = 1$ and 2, it leads to eigenvalues $\lambda_1 = -1$ and $\lambda_2 = -4$, giving $\kappa_1 = 1$ and $\kappa_2 = 2$. Equation (6.4b) has two eigenfunctions, which are the associated Legendre polynomials P_2^1 and P_2^2:

$$P_2^1 = -\frac{1}{2}(1 - s^2)^{\frac{1}{2}}\frac{d}{ds}(3s^2 - 1) = -3s(1 - s^2)^{\frac{1}{2}}$$

and

$$P_2^2 = \frac{1}{2}(1 - s^2)\frac{d^2}{ds^2}(3s^2 - 1) = 3(1 - s^2)$$

Thus, normalized eigenfunctions are obtained as

$$\varphi_1(x, \kappa_1) = \mp\sqrt{\frac{3}{2}} \tanh x \, \mathrm{sech} x$$

and

$$\varphi_2(x, \kappa_2) = \pm\frac{\sqrt{3}}{2} \mathrm{sech}^2 x$$

Apply the asymptotic relation (4.2a), it yields

$$c_1(0) = \lim_{x \to \infty} \varphi_1(x, \kappa_1) e^{\kappa_1 x}$$

$$= \lim_{x \to \infty} \left[\mp\sqrt{\frac{3}{2}} \tanh x \, \mathrm{sech} x e^x \right] = \mp\sqrt{6}$$

and

$$c_2(0) = \lim_{x \to \infty} \varphi_2(x, \kappa_2) e^{\kappa_2 x} = \lim_{x \to \infty} \left[\pm\frac{\sqrt{3}}{2} \mathrm{sech}^2 x e^{2x} \right] = \pm 2\sqrt{3}$$

The kernel function (6.2) is then determined explicitly as

$$K^{(2)}(x, x; t) = -6 \frac{e^{8t-2x} + 2e^{64t-4x} + e^{72t-6x}}{1 + 3e^{8t-2x} + 3e^{64t-4x} + e^{72t-6x}} \tag{6.2a}$$

3. $N = 3$ (three-solitons solution), having the initial condition $\phi^{(3)}(x, 0) = 12 \, \mathrm{sech}^2 x$.

For $N = 3$, $n = 1, 2$, and 3, the eigenvalues are $\lambda_1 = -1, \lambda_2 = -4$, and $\lambda_3 = -9$, which lead to $\kappa_1 = 1$, $\kappa_2 = 2$, and $\kappa_3 = 3$. The three eigenfunctions of (6.4b) are the associated Legendre polynomials P_3^1, P_3^2 and P_3^3:

$$P_3^1 = -\frac{1}{48}(1 - s^2)^{\frac{1}{2}} \frac{d^4}{ds^4}(s^2 - 1)^3 = -\frac{3}{2}(1 - s^2)^{\frac{1}{2}}(5s^2 - 1)$$

$$= -\frac{3}{2}(4 - 5 \, \mathrm{sech}^2 x) \, \mathrm{sech} \, x$$

$$P_3^2 = \frac{1}{48}(1 - s^2) \frac{d^5}{ds^5}(s^2 - 1)^3 = 15(1 - s^2)s = 15 \, \mathrm{sech}^2 x \, \tanh x$$

and

$$P_3^3 = -\frac{1}{48}(1 - s^2)^{\frac{3}{2}}\frac{d^6}{ds^6}(s^2 - 1)^3 = -15(1 - s^2)^{\frac{3}{2}} = -15\operatorname{sech}^3 x$$

Thus, normalized eigenfunctions are obtained as

$$\varphi_1(x, \kappa_1) = \mp\frac{\sqrt{3}}{4}(4 - 5\operatorname{sech}^2 x)\operatorname{sech} x$$

$$\varphi_2(x, \kappa_2) = \pm\frac{\sqrt{15}}{2}\operatorname{sech}^2 x\tanh x$$

and

$$\varphi_3(x, \kappa_3) = \mp\frac{\sqrt{15}}{4}\operatorname{sech}^3 x$$

Apply the asymptotic forms (4.2a), their initial scattering amplitudes $c_n(0)$ are obtained as

$$c_1(0) = \lim_{x\to\infty}\varphi_1(x, \kappa_1)e^{\kappa_1 x}$$

$$= \lim_{x\to\infty}\left[\mp\frac{\sqrt{3}}{4}(4 - 5\operatorname{sech}^2 x)\operatorname{sech} xe^x\right] = \mp2\sqrt{3}$$

$$c_2(0) = \lim_{x\to\infty}\varphi_2(x, \kappa_2)e^{\kappa_2 x}$$

$$= \lim_{x\to\infty}\left[\pm\frac{\sqrt{15}}{2}\operatorname{sech}^2 x\tanh xe^{2x}\right] = \pm2\sqrt{15}$$

and

$$c_3(0) = \lim_{x\to\infty}\varphi_3(x, \kappa_3)e^{\kappa_3 x} = \lim_{x\to\infty}\left[\mp\frac{\sqrt{15}}{4}\operatorname{sech}^3 xe^{3x}\right] = \mp2\sqrt{15}$$

With the aid of these results, (6.3) is expressed explicitly as

$$K^{(3)}(x, x; t) = \frac{\partial}{\partial x}\ln\left|A_{ij}^{(3)}\right| = \frac{\left|A_{ij}^{(3)}\right|_x}{\left|A_{ij}^{(3)}\right|} \tag{6.3a}$$

where

$$\left|A_{ij}^{(3)}\right|_x = -12(e^{8t-2x} + 5e^{64t-4x} + 5e^{72t-6x} + 5e^{216t-6x}$$
$$+ 10e^{224t-8x} + 5e^{280t-10x} + e^{288t-12x})$$
$$\left|A_{ij}^{(3)}\right| = 1 + 6e^{8t-2x} + 15e^{64t-4x} + 10e^{72t-6x} + 10e^{216t-6x}$$
$$+ 15e^{224t-8x} + 6e^{280t-10x} + e^{288t-12x}$$

4. $N = 4$ (four-solitons solution), having the initial condition $\phi^{(4)}(x,0) = 20 \operatorname{sech}^2 x$.

In this case, $n = 1, 2, 3,$ and 4, the eigenvalues are $\lambda_1 = -1$, $\lambda_2 = -4$, $\lambda_3 = -9$, and $\lambda_4 = -16$, giving $\kappa_1 = 1$, $\kappa_2 = 2$, $\kappa_3 = 3$, and $\kappa_4 = 4$. The four eigenfunctions of (6.4b) are the associated Legendre polynomials $P_4^1, P_4^2, P_4^3,$ and P_4^4:

$$P_4^1 = -\frac{1}{384}(1-s^2)^{\frac{1}{2}}\frac{d^5}{ds^5}(s^2-1)^4 = -\frac{5}{2}(1-s^2)^{\frac{1}{2}}s(7s^2-3)$$
$$= -\frac{5}{2}(4 - 7\operatorname{sech}^2 x)\operatorname{sech} x \tanh x$$

$$P_4^2 = \frac{1}{384}(1-s^2)\frac{d^6}{ds^6}(s^2-1)^4 = \frac{15}{2}(1-s^2)(7s^2-1)$$
$$= \frac{15}{2}\operatorname{sech}^2 x(6 - 7\operatorname{sech}^2 x)$$

$$P_4^3 = -\frac{1}{384}(1-s^2)^{\frac{3}{2}}\frac{d^7}{ds^7}(s^2-1)^4 = -105s(1-s^2)^{\frac{3}{2}}$$
$$= -105\tanh x \operatorname{sech}^3 x$$

and

$$P_4^4 = \frac{1}{384}(1-s^2)^2\frac{d^8}{ds^8}(s^2-1)^4 = 105(1-s^2)^2 = 105\operatorname{sech}^4 x$$

The normalization constants (6.4e) of the eigenfunctions are

$$C_4^1 = \pm\left(\frac{25}{4}\int_{-1}^{1}s^2(7s^2-3)^2\,ds\right)^{\frac{1}{2}} = \pm 2\sqrt{5}$$

$$C_4^2 = \pm \left(\frac{225}{4} \int_{-1}^{1} (1 - s^2)(7s^2 - 1)^2 ds \right)^{\frac{1}{2}} = \pm 6\sqrt{5}$$

$$C_4^3 = \pm 105 \left(\int_{-1}^{1} s^2(1 - s^2)^2 \, ds \right)^{\frac{1}{2}} = \pm 4\sqrt{105}$$

and

$$C_4^4 = \pm 105 \left(\int_{-1}^{1} (1 - s^2)^3 \, ds \right)^{\frac{1}{2}} = \pm 12\sqrt{70}$$

Thus, normalized eigenfunctions are obtained as

$$\varphi_1(x, \kappa_1) = \mp \frac{\sqrt{5}}{4}(4 - 7\,\mathrm{sech}^2 x)\,\mathrm{sech}\,x\,\tanh x$$

$$\varphi_2(x, \kappa_2) = \pm \frac{\sqrt{5}}{4}\,\mathrm{sech}^2 x(6 - 7\,\mathrm{sech}^2 x)$$

$$\varphi_3(x, \kappa_3) = \mp \frac{\sqrt{105}}{4}\,\tanh x\,\mathrm{sech}^3 x$$

and

$$\varphi_4(x, \kappa_4) = \pm \frac{\sqrt{35}}{4\sqrt{2}}\,\mathrm{sech}^4 x$$

Apply the asymptotic relation (4.2a), it yields

$$c_1(0) = \lim_{x \to \infty} \varphi_1(x, \kappa_1)e^{\kappa_1 x}$$

$$= \lim_{x \to \infty} \left[\mp \frac{\sqrt{5}}{4}(4 - 7\mathrm{sech}^2 x)\,\mathrm{sech}\,x\,\tanh x e^x \right] = \mp 2\sqrt{5}$$

$$c_2(0) = \varphi_2(x, \kappa_2)e^{\kappa_2 x}$$

$$= \lim_{x \to \infty} \left[\pm \frac{\sqrt{5}}{4}\,\mathrm{sech}^2 x(6 - 7\,\mathrm{sech}^2 x)e^{2x} \right] = \pm 6\sqrt{5}$$

$$c_3(0) = \lim_{x \to \infty} \varphi_3(x, \kappa_3)e^{\kappa_3 x}$$

$$= \lim_{x \to \infty} \left[\mp \frac{\sqrt{105}}{4}\,\tanh x\,\mathrm{sech}^3 x e^{3x} \right] = \mp 2\sqrt{105}$$

and

$$c_4(0) = \lim_{x \to \infty} \varphi_4(x, \kappa_4) e^{\kappa_4 x}$$

$$= \lim_{x \to \infty} \left[\pm \frac{\sqrt{35}}{4\sqrt{2}} \operatorname{sech}^4 x e^{4x} \right] = \pm 2\sqrt{70}$$

Then,

$$\left| A_{ij}^{(4)} \right| = 2e^{-10(x-40t)} [\cosh 10(x - 40t) + 10 \cosh 8(x - 49t)$$

$$+ 45 \cosh 6(x - 56t) + 70 \cosh 4(x - 46t)$$

$$+ 50 \cosh 4(x - 82t) + 35 \cosh 2(x + 56t)$$

$$+ 175 \cosh 2(x - 88t) + 126 \cosh 120t] \tag{6.5}$$

and

$$K^{(4)}(x, x; t) = -20\{ e^{-2(x-4t)} + 9e^{-4(x-16t)}$$

$$+ 3[5e^{-6(x-12t)} + 7e^{-6(x-36t)}]$$

$$+ 14[5e^{-8(x-28t)} + e^{-8(x-64t)}]$$

$$+ 63[e^{-10(x-28t)} + e^{-10(x-52t)}]$$

$$+ 21[e^{-12(x-24t)} + 5e^{-12(x-48t)}]$$

$$+ 7[5e^{-14(x-52t)} + 7e^{-14x}e^{584t}] + 36e^{-16(x-46t)}$$

$$+ 9e^{-18(x-44t)} + e^{-20(x-40t)} \} / |A_{ij}^{(4)}| \tag{6.6}$$

Step 3: Substitute $K(x, x; t)$ into (4.4a) to obtain the soliton solution $\phi(x, t)$ of the KdV equation (4.7).

1. $N = 1$

Substitute (6.1a) into (4.4a), a one-soliton solution is obtained as

$$\phi^{(1)}(x, t) = 2 \operatorname{sech}^2(x - 4t) \tag{6.7}$$

It satisfies the initial condition $\phi^{(1)}(x, 0) = 2 \operatorname{sech}^2 x$

The conservation laws (3.12a) and (3.12b) indicate that the area and energy of the soliton are conserved

$$\int_{-\infty}^{+\infty} \phi^{(1)}(x,t)\,dx = \int_{-\infty}^{+\infty} \phi^{(1)}(x,0)\,dx = 4$$

and

$$\int_{-\infty}^{+\infty} \phi^{(1)2}(x,t)\,dx = \int_{-\infty}^{+\infty} \phi^{(1)2}(x,0)\,dx = \frac{8}{3}$$

2. $N = 2$

Substitute (6.2a) into (4.4a), a two-solitons solution is obtained as

$$\phi^{(2)}(x,t)$$

$$= 24\frac{e^{72t-6x}\left(e^{-64t+4x} + 4e^{-8t+2x} + 6 + 4e^{8t-2x} + e^{64t-4x}\right)}{(1 + 3e^{8t-2x} + 3e^{64t-4x} + e^{72t-6x})^2}$$

$$= 12\left(\frac{3 + 4\cosh(2x - 8t) + \cosh(4x - 64t)}{[3\cosh(x - 28t) + \cosh(3x - 36t)]^2}\right) \qquad (6.8a)$$

At $t = 0$, (6.8a) reduces to the initial condition:

$$\phi^{(2)}(x,0) = 12\left(\frac{3 + 4\cosh 2x + \cosh 4x}{[3\cosh x + \cosh 3x]^2}\right) = 6\operatorname{sech}^2 x$$

Set $y = x - 4t$ and $z = x - 16t$ then, (6.8a) becomes

$$\phi^{(2)}(x,t)$$

$$= \frac{36 + 48\cosh(2z + 24t)}{[3\cosh(z - 12t) + \cosh(3z + 12t)]^2}$$

$$+ \frac{12\cosh(4y - 48t)}{[3\cosh(y - 24t) + \cosh(3y - 24t)]^2}$$

As $t \to \infty$, at y and z, respectively, (6.8a) becomes

$$\phi^{(2)}(x,t \to \infty) = \frac{96e^{2z}}{[3e^{-z} + e^{3z}]^2} + \frac{24e^{-4y}}{[3e^{-y} + e^{-3y}]^2}$$

$$= 8 \operatorname{sech}^2 2 \left(x - 16t - \frac{1}{4} \ln 3 \right)$$

$$+ 2 \operatorname{sech}^2 \left(x - 4t + \frac{1}{2} \ln 3 \right)$$

$$= \sum_{n=1}^{2} 2\kappa_n^2 \operatorname{sech}^2 \kappa_n (x - 4\kappa_n^2 t - x_n) \qquad (6.8b)$$

where $\kappa_1 = 1$, $\kappa_2 = 2$; $x_1 = -\frac{1}{2} \ln 3 = -\frac{1}{2\kappa_1} \ln \frac{c_1^2(0)}{2\kappa_1}$ and $x_2 = \frac{1}{4}$ $\times \ln 3 = \frac{1}{2\kappa_2} \ln \frac{c_2^2(0)}{2\kappa_2}$. It shows that two solitons propagate separately, with the larger one being in the front to propagate faster; the velocity is proportional to the amplitude.

The area and energy of this solitary wave $\phi^{(2)}(x,t)$ can be evaluated directly, with the aid of (4.4), or via the initial pulse function $\phi^{(2)}(x,0)$:

$$\int_{-\infty}^{+\infty} \phi^{(2)}(x,t)\, dx = -2K^{(2)}(-\infty, -\infty, t) = 12$$

where $K^{(2)}(\infty, \infty, t) = 0$ is recognized, and

$$\int_{-\infty}^{+\infty} \phi^{(2)^2}(x,t)\, dx = \int_{-\infty}^{+\infty} \phi^{(2)^2}(x,0)\, dx = 24$$

Exercise 6.2: Show that two separate solitons have area and energy of $(4, \frac{8}{3})$ and $(8, \frac{64}{3})$, respectively.

3. $N = 3$

Substitute (6.3a) into (4.4a), a three-solitons solution is obtained as

$$\phi^{(3)}(x,t) = 24[\cosh(10x - 280t) + 10\cosh(8x - 224t)$$

$$+ 30\cosh(6x - 216t) + 40\cosh(4x - 208t)$$

$$+ 25\cosh(2x - 152t) + 15\cosh(6x - 72t)$$

$$+ 80\cosh(4x - 64t) + 135\cosh(2x - 56t)$$

$$+ 50 \cosh(2x - 8t) + 126]/[\cosh(6x - 144t)$$
$$+ 6 \cosh(4x - 136t) + 15 \cosh(2x - 80t)$$
$$+ 10 \cosh(72t)]^2 \tag{6.9a}$$

At $t = 0$, (6.9a) reduces to the initial condition:

$$\phi^{(3)}(x, 0) = 24[\cosh 10x + 10 \cosh 8x + 45 \cosh 6x$$
$$+ 120 \cosh 4x + 210 \cosh 2x + 126]$$
$$/[\cosh 6x + 6 \cosh 4x + 15 \cosh 2x + 10]^2$$
$$= 12 \operatorname{sech}^2 x$$

Set $y = x - 4t$, $z = x - 16t$, and $w = x - 36t$; as $t \to \infty$, at y, z, and w, respectively, (6.9a) approaches a stationary solution:

$$\phi^{(3)}(x, t \to \infty) = 24$$

$$\times \left[\frac{\cosh(10y - 240t)}{D_1^{(3)}(y, t)} + \frac{40 \cosh(4z - 144t)}{D_2^{(3)}(z, t)} \right.$$

$$\left. + \frac{15 \cosh(6w + 144t)}{D_3^{(3)}(w, t)} \right]$$

$$= 18 \operatorname{sech}^2 3 \left(x - 36t - \frac{1}{6} \ln 10 \right)$$

$$+ 8 \operatorname{sech}^2 2 \left(x - 16t + \frac{1}{4} \ln \frac{5}{3} \right) + 2 \operatorname{sech}^2 \left(x - 4t + \frac{1}{2} \ln 6 \right)$$

$$= \sum_{n=1}^{3} 2\kappa_n^2 \operatorname{sech}^2 \kappa_n (x - 4\kappa_n^2 t - x_n) \tag{6.9b}$$

where

$$D_1^{(3)}(y, t) = [\cosh(6y - 120t) + 6 \cosh(4y - 120t)$$
$$+ 15 \cosh(2y - 72t) + 10 \cosh(72t)]^2$$
$$D_2^{(3)}(z, t) = [\cosh(6z - 48t) + 6 \cosh(4z - 72t)$$
$$+ 15 \cosh(2z - 48t) + 10 \cosh(72t)]^2$$

$$D_3^{(3)}(w,t) = [\cosh(6w + 72t) + 6\cosh(4w + 8t)$$
$$+ 15\cosh(2w - 8t) + 10\cosh(72t)]^2$$

$\kappa_1 = 1, \kappa_2 = 2, \kappa_3 = 3; x_1 = -\frac{1}{2}\ln 6 = -\frac{1}{2\kappa_1}\ln\frac{c_1^2(0)}{2\kappa_1}, x_2 = -\frac{1}{4}\ln\frac{5}{3} = -\frac{1}{2\kappa_2}\ln\frac{c_2^2(0)}{2\kappa_2(2\kappa_2-1)^2}$, and $x_3 = \frac{1}{6}\ln 10 = \frac{1}{2\kappa_3}\ln\frac{c_3^2(0)}{2\kappa_3}$.

It shows that the initial pulse evolves to three solitons, propagating separately in sequence of size with the larger one being in the front.

The area and energy of this solitary wave $\phi^{(3)}(x,t)$ are

$$\int_{-\infty}^{+\infty} \phi^{(3)}(x,t)\,dx = -2K^{(3)}(-\infty, -\infty, t) = 24$$

and

$$\int_{-\infty}^{+\infty} \phi^{(3)^2}(x,t)\,dx = \int_{-\infty}^{+\infty} \phi^{(3)^2}(x,0)\,dx = 96$$

Exercise 6.3: Show that three separate solitons have the area and energy of $\left(4, \frac{8}{3}\right)\left(8, \frac{64}{3}\right)$, and $(12, 72)$, respectively.

4. $N = 4$

Substituting (6.5) into (4.4a), a four-solitons solution is obtained as

$$\phi^{(4)} = 40\frac{Q_4}{\left|\hat{A}_{ij}^{(4)}\right|^2} \tag{6.10a}$$

where

$$Q_4 = \sum_{i=1}^{9} Q_{4i}$$

with

$$Q_{41} = 16260 + 5600\cosh(144t) + 2450\cosh(288t)$$
$$Q_{42} = 17172\cosh 2(x - 4t) + 4410\cosh 2(x - 28t)$$
$$+ 15750\cosh 2(x - 76t) + 4410\cosh 2(x - 148t)$$
$$+ 1575\cosh 2(x - 220t) + 441\cosh 2(x + 116t)$$

$$Q_{43} = 5616 \cosh 4(x - 16t) + 5040 \cosh 4(x - 52t)$$
$$+ 12600 \cosh 4(x - 40t) + 5040 \cosh 4(x - 112t)$$
$$+ 3528 \cosh 4(x - 76t)$$

$$Q_{44} = 490 \cosh 6(x - 12t) + 9996 \cosh 6(x - 36t)$$
$$+ 1750 \cosh 6(x - 84t) + 5103 \cosh 6(x - 76t)$$
$$+ 1225 \cosh 6(x - 60t)$$

$$Q_{45} = 1260 \cosh 8(x - 28t) + 5292 \cosh 8(x - 64t)$$
$$+ 2016 \cosh 8(x - 34t)$$

$$Q_{46} = 630 \cosh 10(x - 28t) + 630 \cosh 10(x - 52t)$$
$$+ 1575 \cosh(10x - 568t) + 225 \cosh(10x - 664t)$$

$$Q_{47} = 56 \cosh 12(x - 24t) + 560 \cosh 12(x - 48t)$$
$$+ 200 \cosh 12(x - 60t)$$

$$Q_{48} = 63 \cosh(14x - 584t) + 90 \cosh 14(x - 52t)$$

$$Q_{49} = 18 \cosh 16(x - 46t) + \cosh 18(x - 44t)$$

and

$$\left| \hat{A}_{ij}^{(4)} \right| = [\cosh 10(x - 40t) + 10 \cosh 8(x - 49t)$$
$$+ 45 \cosh 6(x - 56t) + 70 \cosh 4(x - 46t)$$
$$+ 50 \cosh 4(x - 82t) + 35 \cosh 2(x + 56t)$$
$$+ 175 \cosh 2(x - 88t) + 126 \cosh 120t]$$

At $t = 0$, (6.10a) reduces to the initial condition:

$$\phi^{(4)}(x, 0) = 20 \operatorname{sech}^2 x$$

A plot showing the evolution of the initial pulse in the early stage is presented in Fig. 6.1; in the plot, the amplitude of the initial pulse is normalized by 20 and time "t" is normalized by 64.

As the time increases, the pulse propagates to the right to form a sequential peak, with a larger one located in the front. Since the initial pulse is reflectionless, periodic waves are not generated. These

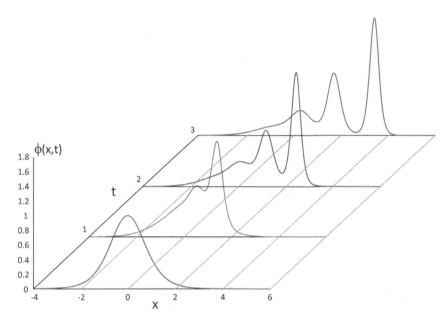

Fig. 6.1. Early-stage evolution of an initial hyperbolic secant square pulse with an amplitude matching four solitons.

peaks will separate to form individual solitons; four solitons are expected.

Set $y = x - 4t$, $z = x - 16t$, $w = x - 36t$, and $u = x - 64t$; as $t \to \infty$, at y, z, w and u, respectively, (6.10a) becomes

$$
\phi^{(4)}(x, t \to \infty) \sim 40 \left\{ \frac{\cosh(18y - 720t)}{[\cosh(10y - 360t) + 10\cosh(8y - 360t)]^2} \right.
$$

$$
+ \frac{200\cosh(12z - 528t)}{[10\cosh(8z - 264t) + 50\cosh(4z - 264t)]^2}
$$

$$
+ \frac{315\cosh(2w - 368t)}{[10\cosh(4w - 184t) + 7\cosh(2w + 184t)]^2}
$$

$$
+ \left. \frac{56\cosh(12u + 480t)}{[\cosh(10u + 240t) + 35\cosh(2u + 240t)]^2} \right\}
$$

$$
\sim 2\,\mathrm{sech}^2 \left(x - 4t + \frac{1}{2}\ln 10 \right)
$$

$$+ 8 \operatorname{sech}^2 \left[2 \left(x - 16t + \frac{1}{4} \ln 5 \right) \right]$$

$$+ 18 \operatorname{sech}^2 \left[3 \left(x - 36t - \frac{1}{6} \ln \frac{10}{7} \right) \right]$$

$$+ 32 \operatorname{sech}^2 \left[4 \left(x - 64t - \frac{1}{8} \ln 35 \right) \right]$$

$$= \sum_{n=1}^{4} 2\kappa_n^2 \operatorname{sech}^2 \kappa_n \left(x - 4\kappa_n^2 t - x_n \right) \qquad (6.10b)$$

where

$$x_1 = -\frac{1}{2} \ln 10 = -\frac{1}{2\kappa_1} \ln \frac{c_1^2(0)}{2\kappa_1}$$

$$x_2 = -\frac{1}{4} \ln 5 = -\frac{1}{2\kappa_2} \ln \frac{c_2^2(0)}{2\kappa_2 \left(2\kappa_2 - 1 \right)^2}$$

$$x_3 = \frac{1}{6} \ln \frac{10}{7} = \frac{1}{2\kappa_3} \ln \frac{c_3^2(0)}{2\kappa_3 \left(2\kappa_3 + 1 \right)^2} \quad \text{and}$$

$$x_4 = \frac{1}{8} \ln 35 = \frac{1}{2\kappa_4} \ln \frac{c_4^2(0)}{2\kappa_4}$$

The area and energy of the solitary wave $\phi^{(4)}(x, t)$ are

$$\int_{-\infty}^{+\infty} \phi^{(4)}(x, t)\, dx = -2K^{(4)}(-\infty, -\infty, t) = 40$$

and

$$\int_{-\infty}^{+\infty} \phi^{(4)^2}(x, t)\, dx = \int_{-\infty}^{+\infty} \phi^{(4)^2}(x, 0)\, dx = \frac{800}{3}$$

Exercise 6.4: Show that four separate solitons have the area and energy of $\left(4, \frac{8}{3} \right)$, $\left(8, \frac{64}{3} \right) (12, 72)$, and $\left(16, \frac{512}{3} \right)$, respectively.

An evolution of the initial pulse to four separated solitons, given by (6.10b), is plotted in Fig. 6.2. As shown, the larger one propagates faster and is in the front; the normalized amplitudes are 1.6, 0.9, 0.4, and 0.1, respectively.

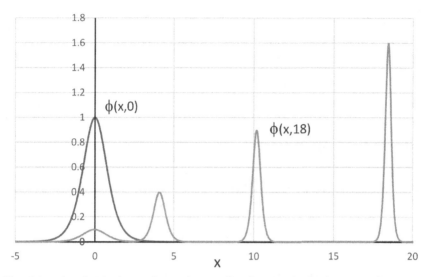

Fig. 6.2. A reflectionless pulse evolves to (four) individual solitons in the steady state.

It is shown that the specific class of the initial pulses in the KdV system evolves into integer number of solitons, traveling at different speeds. The speed of individual soliton is proportional to the amplitude. Larger one locates in the front to move faster.

Exercise 6.5: Explain why the ratio of the four solitons' amplitudes is 1.6:0.9:0.4:0.1.

Ans: Amplitude $\propto \kappa_n^2$.

Exercise 6.6: Explain why in Fig. 6.2 a larger soliton is in the front.

Ans: Soliton velocity $V_n = 4\kappa_n^2$.

6.3 Modeling Gaussian Pulses as Reflectionless Potentials of the Linear Schrödinger Equation (4.6)

Since $\text{sech}^2 x \sim e^{-0.78539x^2}$ an initial Gaussian pulse $\phi(x,0) = Ae^{-\alpha x^2}$ can be modeled as

$$\phi(x,0) = Ae^{-\alpha x^2} = Ae^{-0.78539z^2} \sim A\text{sech}^2 z$$

where $z = 1.128385\sqrt{\alpha}x = ax$ and $a = 1.128385\sqrt{\alpha}$.

Introduce $u(y, \tau) = C\phi(xt)$, $\Phi(y, \tau) = \psi(x, t)$, where $y = px$ and $\tau = qt$; (4.6) and (4.7) become

$$\Phi_{yy} = -\frac{1}{p^2}\left(\frac{1}{C}u + \lambda\right)\Phi \tag{6.11}$$

$$u_\tau + \frac{6p}{Cq}uu_y + \frac{p^3}{q}u_{yyy} = 0 \tag{6.12}$$

Set

$$\frac{p^3}{q} = 1, \quad \frac{p}{Cq} = 1, \quad \frac{A}{p^2} = N(N+1) \quad \text{and} \quad \frac{\lambda}{p^2} = \lambda'$$

where

$$q = p^3, \quad C = \frac{1}{p^2}, \quad CA = N(N+1), \quad \text{and}$$

$$p = \sqrt{\frac{A}{N(N+1)}} = a = 1.128385\sqrt{\alpha}$$

then, (6.11) and (6.12) reduce to

$$\Phi_{yy}(y, 0) \sim -[N(N+1)\text{sech}^2 y + \lambda']\Phi(y, 0) \tag{6.11a}$$

$$u_\tau + 6uu_y + u_{yyy} = 0 \tag{6.12a}$$

subjecting to the initial condition $u(y, 0) \sim N(N+1)\,\text{sech}^2 y$. As shown, (6.11a) can be converted to (6.4b), with $\lambda' = \lambda_n = -\kappa_n^2 = -n^2$, and (6.12a) has the same form and initial condition as (4.7). As $x \to \infty$, (6.11) leads to $\Phi \sim c_n(\tau)\,e^{-\kappa_n y}$ and (4.8b) reduces to

$$\Phi_\tau \sim -4\partial_y^3\Phi \sim 4\kappa_n^3\Phi$$

then, $c_n(\tau) = c_n(0)e^{4\kappa_n^3\tau}$. Thus, (6.12a) has an N-solitons solution $u^{(N)}(y, \tau) = \phi^{(N)}(y, \tau)$; as $t \to \infty$, it is given by

$$u^{(N)}(y, \tau \to \infty) \sim \sum_{n=1}^{N} 2\kappa_n^2\,\text{sech}^2\kappa_n\left(y - 4\kappa_n^2\tau - y_n\right) \tag{6.13}$$

where

$$y_1 = -\frac{1}{2\kappa_1} \ln \frac{c_1^2(0)}{2\kappa_1}; y_N = \frac{1}{2\kappa_N} \ln \frac{c_N^2(0)}{2\kappa_N} \quad \text{for } N > 1$$

and

$$y_n = (-1)^{n+1} \frac{1}{2\kappa_n} \ln \frac{c_n^2(0)}{2\kappa_n \left[2\kappa_n + (-1)^{n+1}\right]^2} \quad \text{for } 1 < n < N$$

and the solution of (4.7) is obtained as

$$\phi_G^{(N)}(x, t \to \infty) \sim \sum_{n=1}^{N} 2a^2 \kappa_n^2 \text{sech}^2 a\kappa_n \left(x - 4a^2 \kappa_n^2 t - x_n\right) \quad (6.13a)$$

where $x_n = \frac{y_n}{a}$.

It is shown that an initial Gaussian pulse $\phi(x,0) = Ae^{-\alpha x^2}$, with the amplitude A and width $1/\alpha$ related by $A = a^2 N(N+1) = 1.27325\alpha N(N+1)$, evolves to N individual solitons. On the other hand, if $\frac{A}{a^2} > N(N+1)$, but close to $N(N+1)$, it evolves to N individual solitons together with linear radiation (i.e., wave with continuous spectrum).

6.4   Pulse Behavior in the Transition Region

The Korteweg–de Vries (KdV) equation

$$\frac{\partial \phi}{\partial t} + \phi \frac{\partial \phi}{\partial x} + 0.001 \frac{\partial^3 \phi}{\partial x^3} = 0 \quad (4.7a)$$

with the initial condition $\phi(x,0) = \exp[-2(5x/3)^2]$ is solved numerically.

The results of Section 6.3 are used to evaluate the number of solitons to appear. It is first to convert (4.7a) to (4.7) with the variable transform: $t' = 2.152t$ and $x' = 12.91x$; the initial condition becomes $\phi(x',0) = (\exp -x'^2/30)$. Thus, $\alpha = 1/30$, $p = 1.128385\sqrt{\alpha} = 0.206$, and $C = 23.56$. For $\text{Max}[N(N+1)] < 23.56$, it leads to $N = 4$.

The numerical results are presented in Fig. 6.3. It demonstrates that this initial Gaussian pulse is indeed decomposed into four

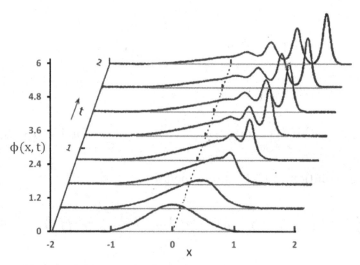

Fig. 6.3.   Numerical solution of KdV equation, showing the evolution of an initial Gaussian pulse.

solitons. It is also consistent with Fig. 6.2, showing that the larger amplitude one is narrower and moves faster; their distance separation increases with time. The transition process illustrates the complexity of such nonlinear wave behavior — breakup to reshape until equilibrium.

Exercise 6.7: If the initial condition of (4.7a) is $\phi(x,0) = \exp[-4(5x/3)^2]$, how many solitons are expected?

Ans: Two ($N = 2$) because $C = 11.78, \text{Max}[N(N+1)] < 11.78$ gives $N = 2$.

6.5   Application of Inverse Scattering Transform (IST): Exemplified with the mKdV and sine-Gordon Equations

6.5.1   *mKdV equation*

The Lax pair $\vec{L}$ and $\vec{A}$ of the mKdV equation (3.15a), via Miura transformation (3.14) on the Lax pair (4.6a) and (4.8c) of the KdV equation, is obtained as

$$\vec{L} = -\frac{\partial^2}{\partial \eta^2} + U_\eta + U^2 \tag{6.14}$$

$$\vec{A} = \gamma - 4\frac{\partial^3}{\partial \eta^3} + 6(U^2 + U_\eta)\frac{\partial}{\partial \eta} + 3(2UU_\eta + U_{\eta\eta}) \tag{6.15}$$

Consider the case that the potential function $V(\eta; \tau) = U_\eta(\eta, \tau) + U^2(\eta, \tau)$ of (6.14) has an initial condition of $V(\eta; 0) = -N(N + 1)\,\text{sech}^2\eta$, then (4.6b), with the aid of (6.14) and setting $s = \tanh\eta$ and $\lambda = -n^2$ for bound states, converts to

$$\frac{d}{ds}(1 - s^2)\frac{d}{ds}\psi + \left[N(N + 1) - \frac{n^2}{(1 - s^2)}\right]\psi = 0$$

which is the same as (6.4b) to obtain the same $c_n(0)$. For bound states, (6.15) leads to

$$c_n(t) = c_n(0)e^{4n^3 t}$$

which is the same as (4.14).

Thus, $V(\eta; \tau)$ is determined, via IST, as given by (6.7) to (6.10a) for $N = 1$ to 4, respectively. It sets up a first-order differential equation for $U(\eta, \tau)$:

$$U_\eta(\eta, \tau) + U^2(\eta, \tau) = V(\eta; \tau) \tag{6.16a}$$

It is solved in the following:

1. For $N = 1$:

With the aid of (6.7),

$$U_\eta(\eta, \tau) + U^2(\eta, \tau) = -2\text{sech}^2(\eta - 4\tau) \tag{6.16b}$$

Rearrange the mKdV equation (3.15a)

$$\frac{\partial}{\partial \tau}U + \frac{\partial^3}{\partial \eta^3}U - 6U^2\frac{\partial}{\partial \eta}U = 0 = \frac{\partial}{\partial \eta}\left[\frac{\partial^2}{\partial \eta^2}U - (4U + 2U^3)\right]$$

$$= \frac{\partial}{\partial \eta}\left[\frac{\partial}{\partial \eta}(U_\eta + U^2) - 2U(U_\eta + U^2) - 4U\right]$$

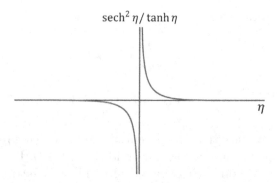

Fig. 6.4.   Representation of a singular solution of the mKdV equation (3.15a).

where $\frac{\partial}{\partial \tau} U = -4 \frac{\partial}{\partial \eta} U$ is applied. It leads to

$$U^{(1)}(\eta, \tau) = \frac{\frac{\partial}{\partial \eta}(U_\eta + U^2)}{2(U_\eta + U^2) + 4} = \frac{\mathrm{sech}^2(\eta - 4\tau)}{\tanh(\eta - 4\tau)} \qquad (6.17a)$$

A representation of this singular solution is plotted in Fig. 6.4.

Exercise 6.8: Show that (6.17a) satisfies the mKdV equation (3.15a).

2. For $N = 2$:

With the aid of (6.8a),

$$U_\eta(\eta, \tau) + U^2(\eta, \tau)$$

$$= -12 \left(\frac{3 + 4\cosh(2\eta - 8\tau) + \cosh(4\eta - 64\tau)}{[3\cosh(\eta - 28\tau) + \cosh(3\eta - 36\tau)]^2} \right) \qquad (6.16c)$$

As $\tau \to \infty$, (6.16c) becomes

$$U_\eta(\eta, \tau \to \infty) + U^2(\eta, \tau \to \infty)$$

$$= -\sum_{n=1}^{2} 2n^2 \mathrm{sech}^2 n(\eta - 4n^2 \tau - x_n)$$

It is solved to obtain

$$U^{(2)}(\eta, \tau \to \infty) = \sum_{n=1}^{2} n \frac{\mathrm{sech}^2 n(\eta - 4n^2 \tau - x_n)}{\tanh n(\eta - 4n^2 \tau - x_n)} \qquad (6.17b)$$

where $x_n = (-1)^n \frac{1}{2n} \ln \frac{c_n^2(0)}{2n}$, i.e., $x_1 = -\frac{1}{2} \ln 3$ and $x_2 = \frac{1}{4} \ln 3$.

Exercise 6.9: Show that $U_n(\eta, \tau) = n \frac{\text{sech}^2 n(\eta - 4n^2 \tau - x_n)}{\tanh n(\eta - 4n^2 \tau - x_n)}$ satisfies the mKdV equation (3.15a).

3. For $N = 3$:

With the aid of (6.9b) for $\tau \to \infty$, (6.16a) becomes

$$U_\eta(\eta, \tau \to \infty) + U^2(\eta, \tau \to \infty) = -\sum_{n=1}^{3} 2n^2 \text{sech}^2 n(\eta - 4n^2 \tau - x_n)$$

which leads to

$$U^{(3)}(\eta, \tau \to \infty) = \sum_{n=1}^{3} n \frac{\text{sech}^2 n(\eta - 4n^2 \tau - x_n)}{n(\eta - 4n^2 \tau - x_n)} \qquad (6.17c)$$

where $x_1 = -\frac{1}{2}\ln 6$, $x_2 = -\frac{1}{4}\ln\frac{5}{3}$ and $x_3 = \frac{1}{6}\ln 10$.

4. For $N = 4$:

With the aid of (6.10b) for $\tau \to \infty$, (6.16a) becomes

$$U_\eta(\eta, \tau \to \infty) + U^2(\eta, \tau \to \infty) = -\sum_{n=1}^{4} 2n^2 \text{sech}^2 n(\eta - 4n^2 \tau - x_n)$$

which is solved to obtain

$$U^{(4)}(\eta, \tau \to \infty) = \sum_{n=1}^{4} n \frac{\text{sech}^2 n(\eta - 4n^2 \tau - x_n)}{\tanh n(\eta - 4n^2 \tau - x_n)} \qquad (6.17d)$$

where $x_1 = -\frac{1}{2}\ln 10$, $x_2 = -\frac{1}{4}\ln 5$, $x_3 = \frac{1}{6}\ln\frac{10}{7}$, and $x_4 = \frac{1}{8}\ln 35$.
The initial conditions $U^{(N)}(\eta, 0)$ for $N = 2$ to 4 are determined numerically by solving

$$U_\eta(\eta, 0) + U^2(\eta, 0) = -N(N+1)\text{sech}^2\eta$$

Then, (6.16a) is solved numerically, with the aid of these initial conditions and $V^{(N)}(\eta; \tau)$ obtained via IST, to obtain N-solitary (singular type) waves solution; moreover, these stationary solutions (6.17b) to (6.17d) are the asymptotic conditions for solving (6.16a) numerically.

6.5.2   *sine-Gordon equation*

$$\varphi_{tx} = \sin \varphi \tag{1.10a}$$

The AKNS pair $[X]$ and $[T]$ are

$$[X] = \begin{bmatrix} -i\lambda & -\dfrac{1}{2}\varphi_x \\[2mm] \dfrac{1}{2}\varphi_x & i\lambda \end{bmatrix} \tag{6.18}$$

and

$$[T] = \frac{i}{4\lambda} \begin{bmatrix} \cos\varphi & \sin\varphi \\ \sin\varphi & -\cos\varphi \end{bmatrix} \tag{6.19}$$

It is noticed that the parameter λ in (6.18) and (6.19) is different from that appearing in (1.10), which is set to be 1 to reduce (1.10) to (1.10a). Set $[\psi] = \begin{bmatrix} \Theta \\ \Psi \end{bmatrix}$, (6.18) and (6.19) give two sets of coupled first-order differential equations:

$$\Theta_x = -i\lambda\Theta - \frac{1}{2}\varphi_x\Psi \tag{6.18a}$$

$$\Psi_x = i\lambda\Psi + \frac{1}{2}\varphi_x\Theta \tag{6.18b}$$

and

$$\Theta_t = \frac{i}{4\lambda}(\cos\varphi\Theta + \sin\varphi\Psi) \tag{6.19a}$$

$$\Psi_t = \frac{i}{4\lambda}(-\cos\varphi\Psi + \sin\varphi\Theta) \tag{6.19b}$$

Combine (6.18a) and (6.18b) into second-order forms

$$\Theta_{xx} + \lambda^2\Theta + \frac{1}{4}\varphi_x^2\Theta + \frac{1}{2}\varphi_{xx}\Psi = 0$$

$$\Psi_{xx} + \lambda^2\Psi + \frac{1}{4}\varphi_x^2\Psi - \frac{1}{2}\varphi_{xx}\Theta = 0$$

It shows that $\Psi = \pm i\Theta$; then,

$$\Theta_{xx} + \lambda^2\Theta + \left(\pm\frac{i}{2}\varphi_{xx} + \frac{1}{4}\varphi_x^2\right)\Theta = 0 \tag{6.20}$$

and (6.19a) becomes

$$\Theta_t = \frac{i}{4\lambda} e^{\pm i\varphi} \Theta \tag{6.21}$$

As $x \to \infty$, $\varphi \to 0$, (6.20) and (6.21) are solved for the bound states (i.e., $\lambda_n = i\kappa_n$) as

$$\Theta_n \sim c_n(t) e^{-\kappa_n x}$$

$$c_n(t) = c_n(0) e^{\frac{1}{4\kappa_n} t} \tag{6.22}$$

Consider a case with the initial condition $\varphi(x,0) = \pm 2i \ln \tanh x$; then, $V(x;0) = -\left[\pm \frac{i}{2}\varphi_{xx}(x,0) + \frac{1}{4}\varphi_x^2(x,0) \right] = -2 \operatorname{sech}^2 x$ and $V(x;t) = -\left[\pm \frac{i}{2}\varphi_{xx}(x,t) + \frac{1}{4}\varphi_x^2(x,t) \right]$ can be determined directly via IST for $N = 1 = n$.

With the aid of (6.22) for $n = 1$, one has

$$\pm \frac{i}{2}\varphi_{xx}(x,t) + \frac{1}{4}\varphi_x^2(x,t) = 2 \operatorname{sech}^2 \left(x - \frac{1}{4}t \right)$$

It is then solved as

$$\varphi(x,t) = \pm i \ln \tanh^2 \left(x - \frac{1}{4}t \right) \tag{6.23a}$$

A plot representing this singular pulse is presented in Fig. 6.5. This singular-type solitary solution is different from those of (1.14b),

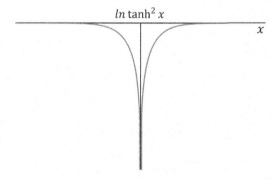

Fig. 6.5. Representation of a singular solution of the sine-Gordon equation (1.10a).

which are kink (+)- and anti-kink (−)-type solutions, plotted in Fig. 1.4.

Set the initial condition $V^{(N)}(x;0) = -\left[\pm \frac{i}{2}\varphi_{xx}(x,0) + \frac{1}{4}\varphi_x^2(x,0)\right] = -N(N+1)\operatorname{sech}^2 x$ and apply a similar IST procedure to that of the previous example, the asymptotic form of an N-solitary (singular type) waves solution is obtained as

$$\varphi^{(N)}(x, t \to \infty) = \pm i \sum_{n=1}^{N} \ln \tanh^2 n \left(x - \frac{1}{4n^2}t - x_n\right) \qquad (6.23b)$$

Exercise 6.10: Show that (6.23a) satisfies the sine-Gordon equation (1.10a).

Problems

P6.1. The initial condition that leads to a pure one-soliton solution for the KdV equation (1.1) is given as $\phi(\eta, 0) = \frac{A}{2}\operatorname{sech}^2\left(\frac{1}{2}\sqrt{A}\eta\right)$ Apply IST to solve (1.1) for the initial condition: $\phi(\eta, 0) = 2\operatorname{sech}^2(\eta)$.

P6.2. In Problem P6.1, the initial condition is changed to $\phi(\eta, 0) = 24\operatorname{sech}^2 2\eta$, find the solution $\phi(\eta, \tau)$ to the KdV equation (1.1).

P6.3. Show that the AKNS equation (4.12), with the AKNS pair $[X]$ and $[T]$ of (6.18) and (6.19), represents the sine-Gordon equation (1.10a).

P6.4. Show that (6.23a) satisfies (1.10a).

Answers to Problems

Chapter 1

P1.1. For the traveling wave solution, $u(x, t) = u(\eta)$ is set, where $\eta = kx - \omega t$ and $\frac{\omega}{k} = V$ is the wave velocity. Then, (P1.1) becomes

$$Dk^2 u_{\eta\eta} + \omega u_\eta + ru(1 - u^q) = 0 \qquad \text{(A1.1)}$$

(1) First, a function $Q(\eta)$ with the property $u = Q^2$ is introduced, where $Q = \left(\frac{1}{1+e^\eta}\right)^{1/q} = \frac{1}{2^{1/q}}\left(1 - \tanh\frac{\eta}{2}\right)^{1/q}$. Then,

$$Q_\eta = -\frac{1}{q}Q(1 - Q^q) \quad \text{and} \quad Q_{\eta\eta} = -\frac{1}{q}Q_\eta[1 - (q+1)Q^q]$$

and

$$u_\eta = 2QQ_\eta, \quad u_{\eta\eta} = -\frac{2}{q}Q[2 - (q+2)Q^q]Q_\eta, \quad \text{and} \quad u(1 - u^q)$$

$$= -qQ(1 + Q^q)Q_\eta$$

With the aid of these relations, (A1.1) reduces to

$$\left(2\omega - \frac{4Dk^2}{q} - rq\right) + \left[\frac{2}{q}(q+2)Dk^2 - rq\right]Q = 0 \qquad \text{(A1.2)}$$

Two equations for the two unknowns k and ω are obtained as

$$2\omega - \frac{4Dk^2}{q} - rq = 0 \quad \text{and} \quad \frac{2}{q}(q+2)Dk^2 - rq = 0 \qquad \text{(A1.3)}$$

141

which lead to $\omega = \frac{rq(q+4)}{2(q+2)}$, $k = \pm\sqrt{\frac{r}{2D(q+2)}}q$, and the wave velocity $V = \sqrt{\frac{rD}{2(q+2)}}(q+4)$.

A solution of (A1.1) is obtained as

$$u = \frac{1}{2^{\frac{2}{q}}}\left(1 - \tanh\frac{\eta}{2}\right)^{\frac{2}{q}}$$

$$= \frac{1}{2^{\frac{2}{q}}}\left[1 \mp \tanh\sqrt{\frac{r}{2D(q+2)}}\frac{q}{2}\left(x \mp \sqrt{\frac{rD}{2(q+2)}}(q+4)t\right)\right]^{\frac{2}{q}}$$

(A1.4)

(2) Fisher equation ($q = 1$):

In this case, $\omega = \frac{5r}{6}$, $k = \pm\sqrt{\frac{r}{6D}}$, and the wave velocity $V = \frac{5\sqrt{rD}}{\sqrt{6}}$; (A1.4) reduces to

$$u(\eta) = \frac{1}{4}\left(1 - \tanh\frac{\eta}{2}\right)^2$$

$$= \frac{1}{4}\left[1 \mp \tanh\frac{1}{2}\sqrt{\frac{r}{6D}}\left(x \mp \frac{5\sqrt{rD}}{\sqrt{6}}t\right)\right]$$

(A1.5)

(3) Burgers–Huxley equation ($D = r = 1$ and $q = 2$):

Thus, $\omega = \frac{3}{2}$, $k = \pm\frac{1}{\sqrt{2}}$, and the wave velocity $V = \frac{3}{\sqrt{2}}$; (A1.4) reduces to

$$u = \frac{1}{2}\left(1 - \tanh\frac{\eta}{2}\right)$$

$$= \frac{1}{2}\left[1 \mp \tanh\frac{1}{2\sqrt{2}}\left(x \mp \frac{3}{\sqrt{2}}t\right)\right]$$

(A1.6)

P1.2. It is first to seek the traveling wave solution by setting $u(x,t) = u(\eta)$, where $\eta = k(x - Vt)$ and V is the wave velocity. Then, (P1.2) becomes

$$(1 - V^2)u_{\eta\eta} + \beta(u^2)_{\eta\eta} + \alpha k^2 u_{\eta\eta\eta\eta} = 0$$

(A1.7)

which is integrated twice as

$$\alpha k^2 u_{\eta\eta} = -\beta u^2 - (1 - V^2)u + C$$

(A1.7a)

where C is an integration constant.

It is now seeking a solution of the form $u(\eta) = P + Q\tanh^2\eta$ and substituting it into (A1.7a), it yields

$$\alpha k^2 Q(1 - 4\tanh^2\eta + 3\tanh^4\eta)$$
$$= C - \beta(P^2 + 2PQ\tanh^2\eta + Q^2\tanh^4\eta)$$
$$- (1 - V^2)(P + Q\tanh^2\eta)$$

Three equations from the constant coefficients of the $\tanh^0\eta(= 1)\tanh^2\eta$, and $\tanh^4\eta$ terms are obtained as

$$\alpha k^2 Q = C - \beta P^2 - (1 - V^2)P$$
$$4\alpha k^2 Q = 2\beta PQ + (1 - V^2)Q$$

and

$$3\alpha k^2 Q = -\beta Q^2$$

Thus,

$$Q = -\frac{3}{\beta}\alpha k^2, \; P = -\frac{1}{2\beta}(1 - V^2 - 4\alpha k^2), \quad \text{and}$$

$$C = -\frac{1}{4\beta}(1 - V^2)^2 + \frac{1}{\beta}(\alpha k^2)^2$$

A solution of (2.4.1) is obtained as

$$u(x,t) = -\frac{1}{2\beta}(1 - V^2 - 4\alpha k^2) - \frac{3}{\beta}\alpha k^2\tanh^2 k(x - Vt)$$

P1.3. Seeking for a traveling wave solution by assuming $u(x,t) = u(\eta)$, where $\eta = x + Vt$. Then, $u_t = Vu_\eta$ and $u_{xxx} = u_{\eta\eta\eta}$, and

(P1.3) becomes

$$-\frac{V}{2}(u^{-2})_\eta = u_{\eta\eta\eta} \tag{A1.8}$$

which is integrated as

$$u^2 u_{\eta\eta} = -\frac{V}{2} = \text{constant}$$

Take a derivative of η on both sides, it yields $u_{\eta\eta\eta} = -\frac{1}{u}(u_\eta^2)_\eta$. Then, (A1.8) becomes

$$(u_\eta^2)_\eta = V\left(\frac{1}{u}\right)_\eta \Rightarrow u_\eta = \sqrt{\frac{V}{u}}$$

It is solved to obtain

$$u(x,t) = [3\alpha(x + 4\alpha^2 t)]^{2/3}$$

where $V = 4\alpha^2$ is set.

P1.4. The coordinate transform leads to

$$\partial_x = -\frac{1}{u}\partial_y \quad \text{and} \quad \partial_t = \partial_\tau + y_t \partial_y$$

Thus,

$$u_x = -2\psi; \; u_{xx} = \frac{2}{u}\psi_y; \; \text{and}$$

$$y_t = \int uu_{xxx}dx = uu_{xx} - \frac{1}{2}u_x^2 = 2(\psi_y - \psi^2)$$

Take x derivative to the Dym equation and apply the Cole–Hopf transformation to it, it yields

$$\psi_t = -6u^2\psi\psi_{xx} + u^3\psi_{xxx}$$

With the aid of

$$\psi_t = \psi_\tau + 2(\psi_y - \psi^2)\psi_y$$

$$\psi_{xx} = \frac{1}{u^2}(\psi_{yy} - 2\psi\psi_y)$$

and

$$\psi_{xxx} = -\frac{1}{u^3}(\psi_{yyy} - 2\psi_y^2 - 6\psi\psi_{yy} + 8\psi^2\psi_y)$$

the Dym equation is transformed into the mKdV equation

$$\psi_\tau + \psi_{yyy} - 6\psi^2\psi_y = 0$$

P1.5.

$$y = -\int \frac{1}{u(x.t)}dx = -\int [3\alpha(x + 4\alpha^2 t)]^{-\frac{2}{3}}dx$$

$$= -\frac{1}{\alpha}[3\alpha(x + 4\alpha^2 t)]^{\frac{1}{3}}$$

It leads to

$$x = -\frac{\alpha^2}{3}(y^3 + 12\tau)$$

Then,

$$\psi = -\frac{1}{2}u_x = -\alpha[3\alpha(x + 4\alpha^2 t)]^{-\frac{1}{3}} = \frac{1}{y}$$

P1.6. The modified Korteweg–de Vries (mKdV) equation for ψ is derived from the Korteweg–de Vries (KdV) equation for ϕ via Miura transformation

$$\phi = -\left(\psi^2 + \frac{\partial}{\partial y}\psi\right)$$

Thus, the solution of the KdV equation is $\phi = 0$, a trivial solution.

P1.7. With the aid of

$$\phi_t = \frac{2}{p} V A_p \mathrm{sech}^{\frac{2}{p}} \xi \tanh \xi$$

$$\phi_x = -\frac{2}{p} A_p \mathrm{sech}^{\frac{2}{p}} \xi \tanh \xi$$

and

$$\phi_{xxx} = -\frac{2}{p} A_p \mathrm{sech}^{\frac{2}{p}} \xi \tanh \xi \left[\left(\frac{2}{p}\right)^2 - \frac{2(p+1)(p+2)}{p^2} \mathrm{sech}^2 \xi \right]$$

(P1.4) becomes

$$\frac{2}{p} A_p \mathrm{sech}^{\frac{2}{p}} \xi \tanh \xi \left[V - \alpha_p A_p^p \mathrm{sech}^2 \xi - \left(\frac{2}{p}\right)^2 \right.$$

$$\left. + \frac{2(p+1)(p+2)}{p^2} \mathrm{sech}^2 \xi \right] = 0$$

It implies that

$$V = \left(\frac{2}{p}\right)^2$$

$$\alpha_p A_p^p = \frac{2(p+1)(p+2)}{p^2}$$

Then, a solitary solution of (P1.4) is obtained as

$$\phi(x,t) = \left[\frac{2(p+1)(p+2)}{\alpha_p p^2} \right]^{\frac{1}{p}} \mathrm{sech}^{\frac{2}{p}} \left(x - \frac{4}{p^2} t \right)$$

P1.8. With the aid of

$$\phi_t = (-iBS + V_s \tanh \xi) A \operatorname{sech} \xi e^{iB\eta}$$
$$\phi_{xx} = [(iB - \tanh \xi)^2 - \operatorname{sech}^2 \xi] A \operatorname{sech} \xi e^{iB\eta}$$

and

$$|\phi|^2 \phi = (A^2 \operatorname{sech}^2 \xi) A \operatorname{sech} \xi e^{iB\eta}$$

(1.25a) becomes

$$\{i(-iBS + V_s \tanh \xi) + [(iB - \tanh \xi)^2 - \operatorname{sech}^2 \xi]$$
$$+ 2A^2 \operatorname{sech}^2 \xi\} A \operatorname{sech} \xi e^{iB\eta} = 0$$

It implies that

$$V_s = 2B$$

and

$$BS - B^2 + \tanh^2 \xi - \operatorname{sech}^2 \xi + 2A^2 \operatorname{sech}^2 \xi = 0$$

which lead to

$$V_s = 2B, \quad A = \pm 1, \quad \text{and} \quad S = \frac{B^2 - 1}{B}$$

Then, (1.25a) has a solitary solution

$$\phi(x, t) = \pm \operatorname{sech} (x - 2Bt) e^{iB\left(x - \frac{B^2 - 1}{B}t\right)}$$

It is the same as (1.32a) with $\kappa = \frac{1}{2}$ and $\beta = -\frac{B}{2}$.

Chapter 2

P2.1. For bounded states, $\varphi_n(\xi, \tau) \sim \varphi_n e^{-\sqrt{2}\alpha_n |\xi|} e^{i\alpha_n^2 \tau}$.
For unbounded states, $\varphi(\xi, \tau) \sim \varphi_0 e^{i\sqrt{2}k|\xi|} e^{-ik^2 \tau}$.

P2.2.

$$-1/2\, \varphi_\ell^* \frac{\partial^2}{\partial \xi_1^2} \varphi_\ell - \alpha_1 |\varphi_\ell|^4 = i\varphi_\ell^* \frac{\partial}{\partial \tau_1} \varphi_\ell$$

$$-1/2\, \varphi_\ell \frac{\partial^2}{\partial \xi_1^2} \varphi_1^* - \alpha_1 |\varphi_\ell|^4 = -i\varphi_\ell \frac{\partial}{\partial \tau_1} \varphi_\ell^*$$

$$i\frac{\partial}{\partial \tau_1} |\varphi_\ell|^2 = -\frac{1}{2}\left(\varphi_\ell^* \frac{\partial^2}{\partial \xi_1^2} \varphi_\ell - \varphi_\ell \frac{\partial^2}{\partial \xi_1^2} \varphi_\ell^*\right)$$

$$= -\frac{1}{2} \frac{\partial}{\partial \xi_1}\left[\varphi_\ell^{*2} \frac{\partial}{\partial \xi_1}\left(\frac{\varphi_\ell}{\varphi_\ell^*}\right)\right]$$

Hence,

$$\frac{d}{d\tau_1} \int_{-\infty}^{\infty} |\varphi_\ell|^2\, d\xi_1 = 0$$

P2.3. Let $x = \eta - 6v\tau$

$$\frac{\partial}{\partial \tau}\tilde{\phi}(\eta, \tau) = \frac{\partial}{\partial \tau}\phi(x, \tau) - 6v\frac{\partial}{\partial x}\phi(x, \tau)$$

$$\frac{\partial}{\partial \eta}\tilde{\phi}(\eta, \tau) = \frac{\partial}{\partial x}\phi(x, \tau) \quad \text{and} \quad \frac{\partial^3}{\partial \eta^3}\tilde{\phi}(\eta, \tau) = \frac{\partial^3}{\partial x^3}\phi(x, \tau)$$

$$\frac{\partial}{\partial \tau}\tilde{\phi} + \frac{\partial^3}{\partial \eta^3}\tilde{\phi} + 6\tilde{\phi}\frac{\partial}{\partial \eta}\tilde{\phi} = \frac{\partial}{\partial \tau}\phi - 6v\frac{\partial}{\partial x}\phi + \frac{\partial^3}{\partial x^3}\phi + 6\phi\frac{\partial}{\partial x}\phi + 6v\frac{\partial}{\partial x}\phi$$

$$= \frac{\partial}{\partial \tau}\phi(x, \tau) + \frac{\partial^3}{\partial x^3}\phi(x, \tau)$$

$$+ 6\phi(x, \tau)\frac{\partial}{\partial x}\phi(x, \tau) = 0$$

P2.4.

$$\phi_1(\eta, \tau) = \frac{a\eta + b}{a\tau + 1}$$

Chapter 3

P3.1. Set $\phi(\xi) = A\,\text{sech}(kx)$; it is found that $E = -\frac{1}{2}k^2$.

P3.2. $z = \sqrt{2}e^{-|\xi|}$

$$\frac{\partial}{\partial \xi} \to -\sqrt{2}e^{-/\xi/}\frac{|\xi|}{\xi}\frac{\partial}{\partial z} \quad \text{and} \quad \frac{\partial^2}{\partial \xi^2} \to z\left[1 - 2\delta(\xi)\right]\frac{\partial}{\partial z} + z^2\frac{\partial^2}{\partial z^2}$$

Hence,

$$\frac{\partial^2}{\partial z^2}y + \frac{1}{z}\frac{\partial}{\partial z}y + \left(1 - \frac{\nu^2}{z^2}\right)y = 0 \quad \text{for} \quad \xi \neq 0$$

P3.3.

$$i\frac{\partial}{\partial \tau}\left(\frac{\partial \varphi^*}{\partial \xi}\frac{\partial \varphi}{\partial \xi}\right) = \frac{\partial \varphi^*}{\partial \xi}\frac{\partial}{\partial \xi}\left(-1/2\frac{\partial^2}{\partial \xi^2}\varphi - \alpha\,|\varphi|^2\,\varphi\right)$$

$$- \frac{\partial \varphi}{\partial \xi}\frac{\partial}{\partial \xi}\left(-1/2\frac{\partial^2}{\partial \xi^2}\varphi^* - \alpha\,|\varphi|^2\,\varphi^*\right)$$

$$= -1/2\frac{\partial}{\partial \xi}\left(\frac{\partial \varphi^*}{\partial \xi}\frac{\partial^2}{\partial \xi^2}\varphi - \frac{\partial \varphi}{\partial \xi}\frac{\partial^2}{\partial \xi^2}\varphi^*\right)$$

$$- \alpha\frac{\partial}{\partial \xi}\,|\varphi|^2\left(\varphi\frac{\partial \varphi^*}{\partial \xi} - \varphi^*\frac{\partial \varphi}{\partial \xi}\right)$$

$$i\frac{\partial}{\partial \tau}\,|\varphi|^4 = 2\,|\varphi|^2\left[\varphi^*\left(-1/2\frac{\partial^2}{\xi^2}\varphi - \alpha\,|\varphi|^2\,\varphi\right)\right.$$

$$\left. - \varphi\left(-1/2\frac{\partial^2}{\partial \xi^2}\varphi^* - \alpha\,|\varphi|^2\,\varphi^*\right)\right]$$

$$= -\,|\varphi|^2\left[\varphi^*\frac{\partial^2}{\partial \xi^2}\varphi - \varphi\frac{\partial^2}{\partial \xi^2}\varphi^*\right]$$

$$= |\varphi|^2\frac{\partial}{\partial \xi}\left(\varphi\frac{\partial \varphi^*}{\partial \xi} - \varphi^*\frac{\partial \varphi}{\partial \xi}\right)$$

Thus,

$$\frac{\partial}{\partial \tau}\left(\left|\frac{\partial \varphi}{\partial \xi}\right|^2 - \alpha\,|\varphi|^4\right) + \frac{\partial}{\partial \xi}\left\{-\frac{i}{2}\left[\left(\frac{\partial \varphi^*}{\partial \xi}\frac{\partial^2}{\partial \xi^2}\varphi - \frac{\partial \varphi}{\partial \xi}\frac{\partial^2}{\xi^2}\varphi^*\right)\right.\right.$$

$$\left.\left. + 2\alpha\,|\varphi|^2\left(\varphi\frac{\partial \varphi^*}{\partial \xi} - \varphi^*\frac{\partial \varphi}{\partial \xi}\right)\right]\right\} = 0$$

Then,

$$\frac{d}{d\tau}H = \frac{i}{4}\int_{-\infty}^{\infty}\frac{\partial}{\partial\xi}\left[\left(\frac{\partial\varphi^*}{\partial\xi}\frac{\partial^2}{\partial\xi^2}\varphi - \frac{\partial\varphi}{\partial\xi}\frac{\partial^2}{\partial\xi^2}\varphi^*\right)\right.$$
$$\left. + 2\alpha\,|\varphi|^2\left(\frac{\partial\varphi^*}{\partial\xi} - \varphi^*\frac{\partial\varphi}{\partial\xi}\right)\right]d\xi = 0$$

P3.4.

$$-1/2E^*\frac{1}{r^2}\frac{d}{dr}r^2\frac{d}{dr}E - \alpha_1|E|^4 = iE^*\frac{\partial}{\partial t}E$$

$$-1/2E\frac{1}{r^2}\frac{d}{dr}r^2\frac{d}{dr}E^* - \alpha_1\,|E|^4 = -iE\frac{\partial}{\partial t}E^*$$

$$i\frac{\partial}{\partial t}|E|^2 = 1/2\left(\frac{E^*}{r^2}\frac{d}{dr}r^2\frac{d}{dr}E - \frac{E}{r^2}\frac{d}{dr}r^2\frac{d}{dr}E^*\right)$$

Let $x = 1/r \Rightarrow \frac{d}{dr} \to -\frac{1}{r^2}\frac{d}{dx}$; thus,

$$i\frac{\partial}{\partial t}|E|^2 = -1/2\frac{1}{r^4}\left(E^*\frac{d^2}{dx^2}E - E\frac{d^2}{dx^2}E^*\right)$$

$$= -1/2\frac{1}{r^4}\frac{\partial}{\partial x}\left[E^{*2}\frac{\partial}{\partial x}\left(\frac{E}{E*}\right)\right]$$

$$\int_0^{\infty}\frac{1}{r^4}\frac{\partial}{\partial x}\left[E^{*2}\frac{\partial}{\partial x}\left(\frac{E}{E^*}\right)\right]r^2dr$$

$$= \int_0^{\infty}\frac{\partial}{\partial x}\left[E^{*2}\frac{\partial}{\partial x}\left(\frac{E}{E^*}\right)\right]dx = 0$$

$$\Rightarrow \frac{d}{dt}\int_0^{2\pi}d\varphi\int_0^{\pi}\sin\theta\,d\theta\int_0^{\infty}|E|^2\,r^2\,dr = 0$$

P3.5.

(1) Consider a plane wave solution $\psi(x,t) = C_1\exp[i(C_2x+C_3-Et)]$. Substitute it into (P3.2), it leads to $E\psi - C_2^2\psi + \alpha C_1^2\psi = 0$, i.e., $E = C_2^2 - \alpha C_1^2$ Then,

$$\psi(x,t) = C_1\exp\{i[C_2x + (\alpha C_1^2 - C_2^2)t + C_3]\}$$

(2) Consider a solution of the form $\psi(x,t) = f_1(t)\exp[if_2(x,t)]$ where $f_1(t) = \frac{C_1}{\sqrt{t}}$ and $f_2(0,t) = \frac{C_2^2}{4t} + (\alpha C_1^2 \ln t + C_3)$; substitute it into (1), it leads to

$$i\psi_t = i\left[\frac{f_{1t}}{f_1} + if_{2t}\right]\psi, \quad \psi_{xx} = \psi[if_{2xx} - f_{2x}^2], \quad \alpha|\psi|^2\psi = \alpha f_1^2\psi$$

Then,

$$f_{2xx} = -\frac{d}{dt}\ln f_1 = \frac{1}{2t} \Rightarrow f_{2x} = \frac{x}{2t} + A(t) \Rightarrow$$

$$f_2 = \frac{x^2}{4t} + A(t)x + B(t)$$

$$f_{2t} + f_{2x}^2 = \alpha f_1^2 = \frac{\alpha C_1^2}{t}$$

$$f_2(0,t) = B(t) = \frac{C_2^2}{4t} + (\alpha C_1^2 \ln t + C_3)$$

$$f_{2t} = \frac{\alpha C_1^2}{t} - \left[\frac{x}{2t} + A(t)\right]^2 = -\frac{x^2}{4t^2} + A_t x - \frac{C_2^2}{4t^2} + \frac{\alpha C_1^2}{t}$$

$$\Rightarrow A(t) = \frac{C_2}{2t}$$

Thus,

$$f_2(x,t) = \frac{x^2}{4t} + A(t)x + B(t) = \frac{x^2}{4t} + \frac{C_2}{2t}x + \frac{C_2^2}{4t}$$

$$+ (\alpha C_1^2 \ln t + C_3)$$

$$\psi(x,t) = \frac{C_1}{\sqrt{t}}\exp\left\{i\left[\frac{(x+C_2)^2}{4t} + (\alpha C_1^2 \ln t + C_3)\right]\right\}$$

P3.6. Set $\psi(x,t) = \exp[i(Bx + Ct + \vartheta)]\phi(Ax - Dt) = \exp[i(Bx + Ct + \vartheta)]\phi(\eta)$, where $\eta = Ax - Dt$, and substitute it into (P3.2) to yield

$$D = 2AB, \text{ i.e., } \eta = Ax - 2ABt, \text{ and}$$

$$\phi_{\eta\eta} + \beta\phi^3 = -E\phi \qquad (\text{A3.1})$$

where $\beta = \frac{\alpha}{A^2}$ and $E = -\frac{C+B^2}{A^2}$.

Equation (A3.1) is integrated into

$$\phi_\eta^2 + E\phi^2 + \frac{\beta}{2}\phi^4 = \text{integration constant}$$

$$= 0 \text{ and then, } \frac{\phi_\eta}{\phi\sqrt{1 + \frac{\beta}{2E}\phi^2}} = \pm\sqrt{-E}$$

It is integrated again to obtain

$$\phi = \pm\sqrt{-\frac{2E}{\beta}}\,\text{sech}(\sqrt{-E}\eta + \theta)$$

(1) Initial condition $\psi(x,0) = \pm C_1\sqrt{\frac{2}{\alpha}}\exp(iC_2)\,\text{sech}(C_1 x + C_3)$ thus,
$B = 0$ and $\vartheta = C_2$; $C = A^2 = C_1^2$, $E = -1$, $\eta = C_1 x$, and

$$\phi(Ax) = \pm C_1\sqrt{\frac{2}{\alpha}}\,\text{sech}(C_1 x + C_3)$$

Then, $\theta = C_3$ and

$$\psi(x,t) = \pm C_1\sqrt{\frac{2}{\alpha}}\frac{\exp[i(C_1^2 t + C_2)]}{\cosh(C_1 x + C_3)}$$

$$= \pm C_1\sqrt{\frac{2}{\alpha}}\exp[i(C_1^2 t + C_2)]\,\text{sech}(C_1 x + C_3)$$

(2) $\psi(x,0) = \pm A\sqrt{\frac{2}{\alpha}}\exp[i(Bx + C_1)]\,\text{sech}(Ax + C_2)$
Compare with the set form $\psi(x,0) = \exp[i(Bx+\vartheta)]\phi(\eta)$, it leads to

$$\vartheta = C_1 \text{ and } \phi(Ax) = \pm A\sqrt{\frac{2}{\alpha}}\,\text{sech}(Ax + C_2)$$

Then, $\sqrt{-E} = 1$, i.e., $C = A^2 - B^2, \theta = C_2, \sqrt{-\frac{2E}{\beta}} = A\sqrt{\frac{2}{\alpha}}$,
and

$$\psi(x,t) = \pm A\sqrt{\frac{2}{\alpha}}\frac{\exp\{i[Bx + (A^2 - B^2)t + C_1]\}}{\cosh(Ax - 2ABt + C_2)}$$

$$= \pm A\sqrt{\frac{2}{\alpha}}\exp\{i[Bx + (A^2 - B^2)t + C_1]\}$$

$$\times \text{sech}(Ax - 2ABt + C_2)$$

P3.7.

(1) $\frac{d}{d\xi} = \frac{ds}{d\xi}\frac{d}{ds} = k\operatorname{sech}^2 k\xi\frac{d}{ds} = k(1-s^2)\frac{d}{ds}$

$$\frac{d^2}{d\xi^2} = k^2(1-s^2)\frac{d}{ds}\left[(1-s^2)\frac{d}{ds}\right]$$

Then, (3.2) is converted into (P3.3).

(2) Convert (P3.3) into (P3.3a), it needs to set

$$2\frac{\alpha\phi^2}{k^2} = N(N+1)(1-s^2) = N(N+1)\operatorname{sech}^2 k\xi$$

$$2\frac{E}{k^2} = -n^2$$

Hence, $\phi \propto \operatorname{sech} k\xi \propto P_1^1$; it leads to $N = n = 1$. Then,

$$k^2 = -2E = 2|E| \text{ and } \phi = \sqrt{\frac{2|E|}{\alpha}}\operatorname{sech}\sqrt{2|E|}\xi$$

P3.8.

(1) From (1.25b), one obtains

$$\Theta\Psi = |\phi|^2,\ \phi^2 = \Theta^2 e^{-2i\lambda(x+2\lambda t)},\ \text{and}\ \phi^{*2} = \Psi^2 e^{2i\lambda(x+2\lambda t)}$$

thus, (1.26) and (1.27) become

$$\begin{bmatrix} \Theta^2 \\ \Psi^2 \end{bmatrix}_x$$

$$= 2i\lambda\begin{bmatrix} -1 & 0 \\ 0 & 1 \end{bmatrix}\begin{bmatrix} \Theta^2 \\ \Psi^2 \end{bmatrix} + 2|\phi|^2\begin{bmatrix} \phi \\ -\phi^* \end{bmatrix}$$

$$= 2\begin{bmatrix} P_1 & 0 \\ 0 & P_2 \end{bmatrix}\begin{bmatrix} \Theta^2 \\ \Psi^2 \end{bmatrix} \tag{A3.2a}$$

where

$$P_1 = -i\lambda + \phi^* e^{-2i\lambda(x+2\lambda t)}$$

$$P_2 = i\lambda + \phi e^{2i\lambda(x+2\lambda t)}$$

and

$$\begin{bmatrix} \Theta^2 \\ \Psi^2 \end{bmatrix}_t$$

$$= 2 \begin{bmatrix} -2i\lambda^2 + i|\phi|^2 & 0 \\ 0 & 2i\lambda^2 - i|\phi|^2 \end{bmatrix} \begin{bmatrix} \Theta^2 \\ \Psi^2 \end{bmatrix} + 2|\phi|^2 \begin{bmatrix} i\phi_x + 2\lambda\phi \\ i\phi_x^* - 2\lambda\phi^* \end{bmatrix}$$

$$= 2 \begin{bmatrix} W_1 & 0 \\ 0 & W_2 \end{bmatrix} \begin{bmatrix} \Theta^2 \\ \Psi^2 \end{bmatrix} \tag{A3.3a}$$

where

$$W_1 = -2i\lambda^2 + i|\phi|^2 + \left(2\lambda + i\frac{\phi_x}{\phi} \right) \phi^* e^{-2i\lambda(x+2\lambda t)}$$

$$W_2 = 2i\lambda^2 - i|\phi|^2 + \left(-2\lambda + i\frac{\phi_x^*}{\phi^*} \right) \phi e^{2i\lambda(x+2\lambda t)}$$

(2)

$$\begin{bmatrix} \phi \\ \phi^* \end{bmatrix}_x = 2i\lambda \begin{bmatrix} -1 & 0 \\ 0 & 1 \end{bmatrix} \begin{bmatrix} \phi \\ \phi^* \end{bmatrix} + |\phi|^2 \begin{bmatrix} e^{-2i\lambda(x+2\lambda t)} \\ e^{2i\lambda(x+2\lambda t)} \end{bmatrix} \tag{A3.2b}$$

$$\begin{bmatrix} \phi \\ \phi^* \end{bmatrix}_t = \begin{bmatrix} -2i\lambda^2 + W_1 & 0 \\ 0 & 2i\lambda^2 + W_2 \end{bmatrix} \begin{bmatrix} \phi \\ \phi^* \end{bmatrix} \tag{A3.3b}$$

(3) From (A3.2b),

$$\phi_x + 2i\lambda\phi = |\phi|^2 e^{-2i\lambda(x+2\lambda t)} \tag{A3.2c}$$

Set $\phi(x,t) = -2\kappa \operatorname{sech} 2\kappa(x + 4\beta t)e^{-i2[\beta x + 2(\beta^2 - \kappa^2)t]}$, then

$$\phi_x + 2i\lambda\phi = 4\kappa \operatorname{sech} 2\kappa(x + 4\beta t)$$
$$\times [\kappa \tanh 2\kappa(x + 4\beta t) + i(\beta - \lambda)]e^{-i2[\beta x + 2(\beta^2 - \kappa^2)t]}$$

and

$$|\phi|^2 e^{-2i\lambda(x+2\lambda t)} = 4\kappa^2 \operatorname{sech}^2 2\kappa(x + 4\beta t)e^{-2i\lambda(x+2\lambda t)}$$

Set $\beta - \lambda = -i\kappa$, then

$$\kappa \tanh 2\kappa(x + 4\beta t) + i(\beta - \lambda) = \kappa[\tanh 2\kappa(x + 4\beta t) + 1]$$
$$= \kappa \operatorname{sech} 2\kappa(x + 4\beta t)e^{2\kappa(x+4\beta t)}$$

Thus,

$$\phi_x + 2i\lambda\phi = 4\kappa^2 \operatorname{sech}^2 2\kappa(x + 4\beta t)e^{-2i\lambda(x+2\lambda t)}$$
$$= |\phi|^2 e^{-2i\lambda(x+2\lambda t)}$$

(4)

$$\begin{bmatrix} \Theta^2 \\ \Psi^2 \end{bmatrix}_{xx} = 2 \begin{bmatrix} 2P_1^2 + P_{1x} & 0 \\ 0 & 2P_2^2 + P_{2x} \end{bmatrix} \begin{bmatrix} \Theta^2 \\ \Psi^2 \end{bmatrix} \tag{A3.4}$$

P3.9. Consider traveling wave solution of the form $U(\eta, \tau) = \varphi(\xi)$, where $\xi = \eta - A_s\tau + \theta$, (3.15a) becomes

$$\frac{d^3}{d\xi^3}\varphi - A_s\frac{d}{d\xi}\varphi - 2\frac{d}{d\xi}\varphi^3 = 0 \tag{A3.5a}$$

This equation is integrated twice to obtain

$$\varphi_\xi = \pm\sqrt{\varphi^4 + A_s\varphi^2 + C_1\varphi + C_0} \tag{A3.5b}$$

In the case of $C_0 = C_1 = 0$, set $y = i\frac{\varphi}{\sqrt{A_s}}$ with $A_s > 0$; then, (A3.5b) reduces to

$$y_\xi = \mp\sqrt{A_s}y\sqrt{1 - y^2}$$

which is integrated as

$$y(\xi) = \pm\operatorname{sech}\sqrt{A_s}\xi$$

Then,

$$U(\eta, \tau) = \varphi(\xi) = -i\sqrt{A_s}y = \mp i\sqrt{A_s}\operatorname{sech}\sqrt{A_s}(\eta - A_s\tau + \theta)$$

P3.10. Apply (3.25)

$$v(x,t) = \frac{1}{2\sqrt{\pi b t}} \int_{-\infty}^{\infty} v(\xi,0) e^{-\frac{(x-\xi)^2}{4bt}} \, d\xi$$

to find $v(x,t)$, and then apply (1.8c)

$$\phi = -2b \frac{v_x}{v}$$

to determine $u(x,t)$.

(1) $v(x,t) = \frac{1}{2\sqrt{\pi b t}} \int_{-\infty}^{\infty} \delta(\xi) e^{-\frac{(x-\xi)^2}{4bt}} \, d\xi = \frac{e^{-\frac{x^2}{4bt}}}{\sqrt{4\pi bt}}$, giving $v_x(x,t) =$

$-\frac{x}{2bt} \frac{e^{-\frac{x^2}{4bt}}}{\sqrt{4\pi bt}}$; then, $\phi(x,t) = \frac{x}{t}$.

(2) $v(x,t) = \frac{1}{2\sqrt{\pi b t}} \int_{-\infty}^{\infty} [1 + \delta(\xi)] e^{-\frac{(x-\xi)^2}{4bt}} \, d\xi = 1 + \frac{e^{-\frac{x^2}{4bt}}}{\sqrt{4\pi bt}}$, giving

$v_x(x,t) = -\frac{x}{2bt} \frac{e^{-\frac{x^2}{4bt}}}{\sqrt{4\pi bt}}$; then, $\phi(x,t) = \frac{x}{t} \dfrac{\frac{e^{-\frac{x^2}{4bt}}}{\sqrt{4\pi bt}}}{1 + \frac{e^{-\frac{x^2}{4bt}}}{\sqrt{4\pi bt}}}$.

(3) $v(x,t) = \frac{V_0}{2\sqrt{\pi b t}} \int_{-\infty}^{\infty} \sin\frac{\pi\xi}{L} e^{-\frac{(x-\xi)^2}{4bt}} \, d\xi = \frac{-iV_0}{4\sqrt{\pi bt}} \int_{-\infty}^{\infty} e^{i\frac{\pi\xi}{L}} e^{-\frac{(x-\xi)^2}{4bt}}$

$\times \, d\xi + \text{c.c.} = \frac{-iV_0}{2} e^{i\frac{\pi x}{L}} e^{-\frac{b\pi^2 t}{L^2}} + \text{c.c.} = V_0 \sin\frac{\pi x}{L} e^{-\frac{b\pi^2 t}{L^2}}$, giving

$v_x(x,t) = \frac{\pi}{L} V_0 \cos\frac{\pi x}{L} e^{-\frac{b\pi^2 t}{L^2}}$; then, $\phi(x,t) = -\frac{2b\pi}{L} \cot\frac{\pi x}{L}$.

Chapter 4

P4.1. For $x \to \pm\infty$, $V(x) \to 0$, then, (4.1a) reduces to

$$\varphi_{xx} + E\varphi = 0$$

Thus, (4.2a) and (4.2b) are solutions of (4.1a) as $x \to \pm\infty$ with $\kappa_n = \sqrt{|E_n|}$ and $k = \sqrt{E}$.

P4.2.

$$\begin{bmatrix} \Theta \\ \Psi \end{bmatrix}_x = \frac{\partial}{\partial x}[\psi] = \begin{bmatrix} -i\lambda & \phi \\ -\phi^* & i\lambda \end{bmatrix} \begin{bmatrix} \Theta \\ \Psi \end{bmatrix}$$

$$= \begin{bmatrix} 0 & \phi \\ -\phi^* & 0 \end{bmatrix}[\psi] - i\lambda \begin{bmatrix} 1 & 0 \\ 0 & -1 \end{bmatrix}[\psi]$$

Then,

$$\left\{ i \begin{bmatrix} 1 & 0 \\ 0 & -1 \end{bmatrix} \frac{\partial}{\partial x} - i \begin{bmatrix} 0 & \phi \\ \phi^* & 0 \end{bmatrix} \right\}[\psi] = \lambda[\psi] = [L][\psi]$$

where

$$[L] = i \begin{bmatrix} 1 & 0 \\ 0 & -1 \end{bmatrix} \frac{\partial}{\partial x} - i \begin{bmatrix} 0 & \phi \\ \phi^* & 0 \end{bmatrix} = i \begin{bmatrix} \dfrac{\partial}{\partial x} & -\phi \\ -\phi^* & -\dfrac{\partial}{\partial x} \end{bmatrix}$$

$$\begin{bmatrix} \Theta \\ \Psi \end{bmatrix}_t = \frac{\partial}{\partial t}[\psi] = \begin{bmatrix} -2i\lambda^2 + i|\phi|^2 & i\phi_x + 2\lambda\phi \\ i\phi_x^* - 2\lambda\phi^* & 2i\lambda^2 - i|\phi|^2 \end{bmatrix} \begin{bmatrix} \Theta \\ \Psi \end{bmatrix}$$

$$= \begin{bmatrix} i|\phi|^2 & i\phi_x \\ i\phi_x^* & -i|\phi|^2 \end{bmatrix}[\psi] + 2 \begin{bmatrix} 0 & \phi \\ -\phi^* & 0 \end{bmatrix} \vec{L}[\psi] - 2i \begin{bmatrix} 1 & 0 \\ 0 & -1 \end{bmatrix} \vec{L}\vec{L}[\psi]$$

$$= [A][\psi]$$

where

$$[A] = \begin{bmatrix} i|\phi|^2 & i\phi_x \\ i\phi_x^* & -i|\phi|^2 \end{bmatrix} + 2i \begin{bmatrix} 0 & \phi \\ -\phi^* & 0 \end{bmatrix} \begin{bmatrix} \dfrac{\partial}{\partial x} & -\phi \\ -\phi^* & -\dfrac{\partial}{\partial x} \end{bmatrix}$$

$$+ 2i \begin{bmatrix} 1 & 0 \\ 0 & -1 \end{bmatrix} \begin{bmatrix} \dfrac{\partial}{\partial x} & -\phi \\ -\phi^* & -\dfrac{\partial}{\partial x} \end{bmatrix} \begin{bmatrix} \dfrac{\partial}{\partial x} & -\phi \\ -\phi^* & -\dfrac{\partial}{\partial x} \end{bmatrix}$$

$$= i \begin{bmatrix} |\phi|^2 & \phi_x \\ \phi_x^* & -|\phi|^2 \end{bmatrix} - 2i \begin{bmatrix} |\phi|^2 & \phi\dfrac{\partial}{\partial x} \\ \phi^*\dfrac{\partial}{\partial x} & -|\phi|^2 \end{bmatrix}$$

$$+ 2i \begin{bmatrix} \dfrac{\partial^2}{\partial x^2} + |\phi|^2 & -\phi_x \\ -\phi_x^* & -\left(\dfrac{\partial^2}{\partial x^2} + |\phi|^2\right) \end{bmatrix}$$

$$= 2i \begin{bmatrix} \dfrac{\partial^2}{\partial x^2} + \dfrac{|\phi|^2}{2} & -\left(\phi\dfrac{\partial}{\partial x} + \dfrac{\phi_x}{2}\right) \\ -\left(\phi^*\dfrac{\partial}{\partial x} + \dfrac{\phi_x^*}{2}\right) & -\left(\dfrac{\partial^2}{\partial x^2} + \dfrac{|\phi|^2}{2}\right) \end{bmatrix}$$

$$= 2i \begin{bmatrix} 1 & 0 \\ 0 & -1 \end{bmatrix} \dfrac{\partial^2}{\partial x^2} - 2i \begin{bmatrix} 0 & \phi \\ \phi^* & 0 \end{bmatrix} \dfrac{\partial}{\partial x} + i \begin{bmatrix} |\phi|^2 & -\phi_x \\ -\phi_x^* & -|\phi|^2 \end{bmatrix}$$

P4.3.

$$[L] = i \begin{bmatrix} \dfrac{\partial}{\partial x} & -\phi \\ -\phi^* & -\dfrac{\partial}{\partial x} \end{bmatrix} \Rightarrow [L]_t == i \begin{bmatrix} 0 & -\phi_t \\ -\phi_t^* & 0 \end{bmatrix}$$

$$\{[L], [A]\} = [L][A] - [A][L]$$

$$= -2 \begin{bmatrix} \dfrac{\partial}{\partial x} & -\phi \\ -\phi^* & -\dfrac{\partial}{\partial x} \end{bmatrix} \begin{bmatrix} \dfrac{\partial^2}{\partial x^2} + \dfrac{|\phi|^2}{2} & -\left(\phi\dfrac{\partial}{\partial x} + \dfrac{\phi_x}{2}\right) \\ -\left(\phi^*\dfrac{\partial}{\partial x} + \dfrac{\phi_x^*}{2}\right) & -\left(\dfrac{\partial^2}{\partial x^2} + \dfrac{|\phi|^2}{2}\right) \end{bmatrix}$$

$$+ 2 \begin{bmatrix} \dfrac{\partial^2}{\partial x^2} + \dfrac{|\phi|^2}{2} & -\left(\phi\dfrac{\partial}{\partial x} + \dfrac{\phi_x}{2}\right) \\ -\left(\phi^*\dfrac{\partial}{\partial x} + \dfrac{\phi_x^*}{2}\right) & -\left(\dfrac{\partial^2}{\partial x^2} + \dfrac{|\phi|^2}{2}\right) \end{bmatrix} \begin{bmatrix} \dfrac{\partial}{\partial x} & -\phi \\ -\phi^* & -\dfrac{\partial}{\partial x} \end{bmatrix}$$

$$
= - \begin{bmatrix} 2\dfrac{\partial^3}{\partial x^3} + 3|\phi|^2\dfrac{\partial}{\partial x} + |\phi|_x^2 + \phi\phi_x^* & -3\phi_x\dfrac{\partial}{\partial x} - \phi_{xx} + \phi|\phi|^2 \\[2ex] 3\phi_x^*\dfrac{\partial}{\partial x} + \phi_{xx}^* - \phi^*|\phi|^2 & 2\dfrac{\partial^3}{\partial x^3} + 3|\phi|^2\dfrac{\partial}{\partial x} + |\phi|_x^2 + \phi^*\phi_x \end{bmatrix}
$$

$$
+ \begin{bmatrix} 2\dfrac{\partial^3}{\partial x^3} + 3|\phi|^2\dfrac{\partial}{\partial x} + |\phi|_x^2 + \phi\phi_x^* & -3\phi_x\dfrac{\partial}{\partial x} - 2\phi_{xx} - \phi|\phi|^2 \\[2ex] 3\phi_x^*\dfrac{\partial}{\partial x} + 2\phi_{xx}^* + \phi^*|\phi|^2 & 2\dfrac{\partial^3}{\partial x^3} + 3|\phi|^2\dfrac{\partial}{\partial x} + |\phi|_x^2 + \phi^*\phi_x \end{bmatrix}
$$

$$
= \begin{bmatrix} 0 & -\phi_{xx} - 2\phi|\phi|^2 \\[2ex] \phi_{xx}^* + 2\phi^*|\phi|^2 & 0 \end{bmatrix}
$$

Thus,

$$
i\begin{bmatrix} 0 & -\phi_t \\ -\phi_t^* & 0 \end{bmatrix} + \begin{bmatrix} 0 & -\phi_{xx} - 2\phi|\phi|^2 \\ \phi_{xx}^* + 2\phi^*|\phi|^2 & 0 \end{bmatrix} = 0
$$

$$
= \begin{bmatrix} 0 & i\phi_t + \phi_{xx} + 2\phi|\phi|^2 \\ i\phi_t^* - \phi_{xx}^* - 2\phi^*|\phi|^2 & 0 \end{bmatrix}
$$

$$
\Rightarrow i\phi_t + \phi_{xx} + 2\phi|\phi|^2 = 0
$$

P4.4. From (1.26) and (1.27),

$$
\begin{bmatrix} \Theta \\ \Psi \end{bmatrix}_x = \begin{bmatrix} -i\lambda & \phi \\ -\phi^* & i\lambda \end{bmatrix}\begin{bmatrix} \Theta \\ \Psi \end{bmatrix} \tag{1.26}
$$

$$
\begin{bmatrix} \Theta \\ \Psi \end{bmatrix}_t = \begin{bmatrix} -2i\lambda^2 + i|\phi|^2 & i\phi_x + 2\lambda\phi \\ i\phi_x^* - 2\lambda\phi^* & 2i\lambda^2 - i|\phi|^2 \end{bmatrix}\begin{bmatrix} \Theta \\ \Psi \end{bmatrix} \tag{1.27}
$$

Thus,

$$
[X] = \begin{bmatrix} -i\lambda & \phi \\ -\phi^* & i\lambda \end{bmatrix}
$$

and

$$
[T] = \begin{bmatrix} -2i\lambda^2 + i|\phi|^2 & i\phi_x + 2\lambda\phi \\ i\phi_x^* - 2\lambda\phi^* & 2i\lambda^2 - i|\phi|^2 \end{bmatrix}
$$

Substitute these operators into the AKNS equation

$$[X]_t - [T]_x + \{[X], [T]\} = 0$$

it yields

$$\begin{bmatrix} 0 & \phi_t \\ -\phi_t^* & 0 \end{bmatrix} - \begin{bmatrix} i|\phi|_x^2 & i\phi_{xx} + 2\lambda\phi_x \\ i\phi_{xx}^* - 2\lambda\phi_x^* & -i|\phi|_x^2 \end{bmatrix}$$

$$+ \begin{bmatrix} i|\phi|_x^2 & 2\lambda\phi_x - 2i|\phi|^2\phi \\ -2\lambda\phi_x^* - 2i\phi^*|\phi|^2 & -i|\phi|_x^2 \end{bmatrix}$$

$$\begin{bmatrix} 0 & \phi_t - i(\phi_{xx} + 2|\phi|^2\phi) \\ -\phi_t^* - i(\phi_{xx}^* + 2|\phi|^2\phi^*) & 0 \end{bmatrix} = 0$$

It implies (1.25a).

Via AKNS equation:

Set $X_{11} = -i\lambda = -X_{22}$, $X_{12} = \phi$, and $X_{21} = -\phi^*$, i.e.,

$$[X] = \begin{bmatrix} -i\lambda & \phi \\ -\phi^* & i\lambda \end{bmatrix}$$

so that the determinant of the matrix $[I]\partial_x - [X]$ equals to $\partial_x^2 + \lambda^2 + |\phi|^2$, the same as that of the matrix $[L] - \lambda[I]$ from P4.3, where $[I] = \begin{bmatrix} 1 & 0 \\ 0 & 1 \end{bmatrix}$, an identity matrix; then,

$$[X]_t = \begin{bmatrix} 0 & \phi_t \\ -\phi_t^* & 0 \end{bmatrix} = \begin{bmatrix} 0 & i(\phi_{xx} + 2|\phi|^2\phi) \\ i\left(\phi_{xx}^* + 2|\phi|^2\phi^*\right) & 0 \end{bmatrix}$$

Set $T_{11} = -T_{22}$, i.e., $[T] = \begin{bmatrix} T_{11} & T_{12} \\ T_{21} & -T_{11} \end{bmatrix}$; then,

$$[T]_x = \begin{bmatrix} T_{11x} & T_{12x} \\ T_{21x} & -T_{11x} \end{bmatrix}$$

$$[X][T] - [T][X] = \left\{ \begin{bmatrix} -i\lambda & \phi \\ -\phi^* & i\lambda \end{bmatrix} \begin{bmatrix} T_{11} & T_{12} \\ T_{21} & -T_{11} \end{bmatrix} \right.$$

$$- \begin{bmatrix} T_{11} & T_{12} \\ T_{21} & -T_{11} \end{bmatrix} \begin{bmatrix} -i\lambda & \phi \\ -\phi^* & i\lambda \end{bmatrix} \right\}$$

$$= \left\{ \begin{bmatrix} -i\lambda T_{11} + \phi T_{21} & -i\lambda T_{12} - \phi T_{11} \\ i\lambda T_{21} - \phi^* T_{11} & -i\lambda T_{11} - \phi^* T_{12} \end{bmatrix} \right.$$

$$- \begin{bmatrix} -i\lambda T_{11} - \phi^* T_{12} & i\lambda T_{12} + T_{11}\phi \\ -i\lambda T_{21} + T_{11}\phi^* & -i\lambda T_{11} + T_{21}\phi \end{bmatrix} \right\}$$

$$= \begin{bmatrix} \phi T_{21} + \phi^* T_{12} & -2(i\lambda T_{12} + T_{11}\phi) \\ 2(i\lambda T_{21} - \phi^* T_{11}) & -(\phi T_{21} + \phi^* T_{12}) \end{bmatrix}$$

The AKNS equation

$$[X]_t - [T]_x + \{[X], [T]\} = 0$$

leads to

$$\begin{bmatrix} 0 & i(\phi_{xx} + 2|\phi|^2 \phi) \\ i(\phi_{xx}^* + 2|\phi|^2 \phi^*) & 0 \end{bmatrix} - \begin{bmatrix} T_{11x} & T_{12x} \\ T_{21x} & -T_{11x} \end{bmatrix}$$

$$+ \begin{bmatrix} \phi T_{21} + \phi^* T_{12} & -2(i\lambda T_{12} + T_{11}\phi) \\ 2(i\lambda T_{21} - \phi^* T_{11}) & -(\phi T_{21} + \phi^* T_{12}) \end{bmatrix} = 0$$

It implies that

$$T_{11x} = \phi T_{21} + \phi^* T_{12}$$
$$T_{12x} + 2(i\lambda T_{12} + T_{11}\phi) = i(\phi_{xx} + 2|\phi|^2 \phi)$$
$$T_{21x} - 2(i\lambda T_{21} - \phi^* T_{11}) = i(\phi_{xx}^* + 2|\phi|^2 \phi^*)$$

These relations lead to

$$T_{12} = i\phi_x + 2\lambda\phi, \ T_{21} = i\phi_x^* - 2\lambda\phi^*, \text{ and}$$
$$T_{11} = -2i\lambda^2 + i\,|\phi|^2$$

The matrix $[T]$ is then determined as

$$[T] = \begin{bmatrix} -2i\lambda^2 + i\,|\phi|^2 & i\phi_x + 2\lambda\phi \\ i\phi_x^* - 2\lambda\phi^* & 2i\lambda^2 - i\,|\phi|^2 \end{bmatrix}$$

P4.5. Modified KdV (mKdV) equation (3.15a)

$$\frac{\partial}{\partial\tau}U + \frac{\partial^3}{\partial\eta^3}U - 6U^2\frac{\partial}{\partial\eta}U = 0$$

can be obtained from the KdV equation (1.1)

$$\frac{\partial}{\partial t}\phi + \frac{\partial^3}{\partial x^3}\phi + 6\phi\frac{\partial}{\partial x}\phi = 0$$

via Miura transformation (3.14)

$$\phi(\eta, \tau) = -(U^2 + U_\eta)$$

Thus, the Lax pair $\vec{L}$ and $\vec{A}$ of the mKdV equation can also be obtained from the Lax pair (4.6a) and (4.8c) of the KdV equation via Miura transformation (3.14). It results to

$$\vec{L} = -\frac{\partial^2}{\partial\eta^2} + U_\eta + U^2$$

$$\vec{A} = \gamma - 4\frac{\partial^3}{\partial\eta^3} + 6(U^2 + U_\eta)\frac{\partial}{\partial\eta} + 3(2UU_\eta + U_{\eta\eta})$$

Thus, in the Lax equation (4.9),

$$\vec{L}_\tau = U_{\eta\tau} + 2UU_\tau$$

$$[\vec{L}, \vec{A}] = \vec{L}\vec{A} - \vec{A}\vec{L} = \left(\frac{\partial^2}{\partial\eta^2} - U_\eta - U^2\right)$$

$$\times \left[4\frac{\partial^3}{\partial\eta^3} - 6(U^2 + U_\eta)\frac{\partial}{\partial\eta} - 3(2UU_\eta + U_{\eta\eta})\right]$$

$$- \left[4\frac{\partial^3}{\partial\eta^3} - 6(U^2 + U_\eta)\frac{\partial}{\partial\eta} - 3(2UU_\eta + U_{\eta\eta})\right]$$

$$\times \left(\frac{\partial^2}{\partial\eta^2} - U_\eta - U^2\right)$$

$$= -\left\{\frac{\partial^2}{\partial\eta^2}\left[6\left(U^2 + U_\eta\right)\frac{\partial}{\partial\eta} + 3(2UU_\eta + U_{\eta\eta})\right]\right.$$

$$\left. - \left[6(U^2 + U_\eta)\frac{\partial}{\partial\eta} + 3(2UU_\eta + U_{\eta\eta})\right]\frac{\partial^2}{\partial\eta^2}\right\}$$

$$- \left\{(U_\eta + U^2)\left[4\frac{\partial^3}{\partial\eta^3} - 6(U^2 + U_\eta)\frac{\partial}{\partial\eta}\right]\right.$$

$$\left. - \left[4\frac{\partial^3}{\partial\eta^3} - 6(U^2 + U_\eta)\frac{\partial}{\partial\eta}\right](U_\eta + U^2)\right\}$$

where

$$\frac{\partial^2}{\partial\eta^2}\left[6(U^2 + U_\eta)\frac{\partial}{\partial\eta} + 3(2UU_\eta + U_{\eta\eta})\right]$$

$$= 6\left[(U^2 + U_\eta)\frac{\partial^3}{\partial\eta^3} + 2(2UU_\eta + U_{\eta\eta})\frac{\partial^2}{\partial\eta^2}\right.$$

$$\left. + (2U_\eta^2 + 2UU_{\eta\eta} + U_{\eta\eta\eta})\frac{\partial}{\partial\eta}\right]$$

$$+ 3\left[(2UU_\eta + U_{\eta\eta})\frac{\partial^2}{\partial\eta^2} + 2(2U_\eta^2 + 2UU_{\eta\eta} + U_{\eta\eta\eta})\frac{\partial}{\partial\eta}\right.$$

$$\left. + (6U_\eta U_{\eta\eta} + 2UU_{\eta\eta\eta} + U_{\eta\eta\eta\eta})\right]$$

and

$$\left[4\frac{\partial^3}{\partial\eta^3} - 6(U^2 + U_\eta)\frac{\partial}{\partial\eta}\right](U_\eta + U^2)$$

$$= 4\left[(U_\eta + U^2)\frac{\partial^3}{\partial\eta^3} + 3(2UU_\eta + U_{\eta\eta})\frac{\partial^2}{\partial\eta^2}\right.$$

$$\left. + 3(2U_\eta^2 + 2UU_{\eta\eta} + U_{\eta\eta\eta})\frac{\partial}{\partial\eta} + (6U_\eta U_{\eta\eta} + 2UU_{\eta\eta\eta} + U_{\eta\eta\eta\eta})\right]$$

$$- 6\left[(U^2 + U_\eta)^2\frac{\partial}{\partial\eta} + (U^2 + U_\eta)(2UU_\eta + U_{\eta\eta})\right]$$

Then,

$$[\vec{L}, \vec{A}] = -6\left[2(2UU_\eta + U_{\eta\eta})\frac{\partial^2}{\partial\eta^2} + (2U_\eta^2 + 2UU_{\eta\eta} + U_{\eta\eta\eta})\frac{\partial}{\partial\eta}\right]$$

$$- 3\left[2(2U_\eta^2 + 2UU_{\eta\eta} + U_{\eta\eta\eta})\frac{\partial}{\partial\eta}\right.$$

$$\left. + (6U_\eta U_{\eta\eta} + 2UU_{\eta\eta\eta} + U_{\eta\eta\eta\eta})\right]$$

$$+ 4\left[3(2UU_\eta + U_{\eta\eta})\frac{\partial^2}{\partial\eta^2} + 3(2U_\eta^2 + 2UU_{\eta\eta} + U_{\eta\eta\eta})\frac{\partial}{\partial\eta}\right.$$

$$\left. + (6U_\eta U_{\eta\eta} + 2UU_{\eta\eta\eta} + U_{\eta\eta\eta\eta})\right] - 6[(U^2 + U_\eta)(2UU_\eta + U_{\eta\eta})]$$

$$= (6U_\eta U_{\eta\eta} + 2UU_{\eta\eta\eta} + U_{\eta\eta\eta\eta}) - 6\left[(U^2 + U_\eta)(2UU_\eta + U_{\eta\eta})\right]$$

$$= 2UU_{\eta\eta\eta} + U_{\eta\eta\eta\eta} - 12U^3U_\eta - 6(U^2U_\eta)_\eta$$

The Lax equation leads to

$$U_{\eta\tau} + 2UU_\tau + 2UU_{\eta\eta\eta} + U_{\eta\eta\eta\eta} - 12U^3U_\eta - 6(U^2U_\eta)_\eta$$

$$= \left(\frac{\partial}{\partial\eta} + 2U\right)(U_\tau - 6U^2U_\eta + U_{\eta\eta\eta}) = 0$$

It implies that

$$U_\tau - 6U^2U_\eta + U_{\eta\eta\eta} = 0$$

P4.6.

Solution 1:

Apply Miura transformation (3.14) to (4.6d) and (4.8e) of the KdV Lax pair, it results to

$$[L] = i \begin{bmatrix} \partial_\eta & (U^2 + U_\eta) \\ -1 & -\partial_\eta \end{bmatrix}$$

$$[A] = \begin{bmatrix} a_{11} & a_{12} \\ a_{21} & a_{22} \end{bmatrix}$$

where

$$a_{11} = a_{22} = \gamma - 4\partial_\eta^3 + 6(U^2 + U_\eta)\partial_\eta + 3(2UU_\eta + U_{\eta\eta})$$
$$a_{12} = -6(2UU_\eta + U_{\eta\eta})\partial_\eta - 3(2U_\eta^2 + 2UU_{\eta\eta} + U_{\eta\eta\eta})$$
$$a_{21} = 0$$

With the defined matrix Lax pair, (4.6c) and (4.8d) lead to the same equations $\vec{L}\psi = \lambda\psi$ and $\psi_t = \vec{A}\psi$, given in P4.5 (with ψ and λ replaced by η and λ^2); accordingly, the matrix Lax equation (4.9a) of the defined matrix Lax pair represents the mKdV equation.

Solution 2:

The Lax pair [L] and [A] can also be determined with symmetric forms.

Introduce

$$\vec{L} = i \begin{bmatrix} 1 & 0 \\ 0 & -1 \end{bmatrix} \partial_x + \begin{bmatrix} 0 & u \\ u & 0 \end{bmatrix} = -i \begin{bmatrix} -\partial_x & iu \\ iu & \partial_x \end{bmatrix}$$

and

$$\vec{A} = \begin{bmatrix} f & g \\ -g & f \end{bmatrix}$$

Hence,

$$\vec{L}_t = \begin{bmatrix} 0 & u_t \\ u_t & 0 \end{bmatrix} = \begin{bmatrix} 0 & 6u^2 u_x - u_{xxx} \\ 6u^2 u_x - u_{xxx} & 0 \end{bmatrix}$$

$$\left[\vec{L}, \vec{A}\right] = -i\left\{ \begin{bmatrix} -\partial_x & iu \\ iu & \partial_x \end{bmatrix} \begin{bmatrix} f & g \\ -g & f \end{bmatrix} - \begin{bmatrix} f & g \\ -g & f \end{bmatrix} \begin{bmatrix} -\partial_x & iu \\ iu & \partial_x \end{bmatrix} \right\}$$

$$= -i\left\{ \begin{bmatrix} -\partial_x f - iug & -\partial_x g + iuf \\ -\partial_x g + iuf & \partial_x f + iug \end{bmatrix} - \begin{bmatrix} -f\partial_x + igu & g\partial_x + ifu \\ g\partial_x + ifu & f\partial_x - igu \end{bmatrix} \right\}$$

$$= -i\begin{bmatrix} -(\partial_x f - f\partial_x) - i(ug + gu) & -(\partial_x g + g\partial_x) + i(uf - fu) \\ -(\partial_x g + g\partial_x) + i(uf - fu) & (\partial_x f - f\partial_x) + i(ug + gu) \end{bmatrix}$$

Substitute these into the Lax equation $\vec{L}_t + \left[\vec{L}, \vec{A}\right] = 0$, it leads to

$$(\partial_x f - f\partial_x) + i(ug + gu) = 0$$

$$(\partial_x g + g\partial_x) - i(uf - fu) = i(6u^2 u_x - u_{xxx})$$

We now try

$$f = a_0 + a_1\partial_x + a_2\partial_x^2 + a_3\partial_x^3 \quad \text{and} \quad g = b_0 + b_1\partial_x + b_2\partial_x^2$$

Then,

$$(\partial_x f - f\partial_x) + i(ug + gu) = a_{0x} + a_{1x}\partial_x + a_{2x}\partial_x^2 + a_{3x}\partial_x^3$$
$$+ i[2u(b_0 + b_1\partial_x + b_2\partial_x^2) + b_1 u_x + b_2(2u_x\partial_x + u_{xx})]$$
$$= [a_{0x} + i(2ub_0 + b_1 u_x + b_2 u_{xx})] + [a_{1x} + i2(ub_1 + u_x b_2)]\partial_x$$
$$+ (a_{2x} + i2ub_2)\partial_x^2 + a_{3x}\partial_x^3 = 0$$

and

$$(\partial_x g + g\partial_x) - i(uf - fu)$$
$$= 2(b_0 + b_1\partial_x + b_2\partial_x^2)\partial_x + b_{0x} + b_{1x}\partial_x + b_{2x}\partial_x^2$$
$$+ i[a_1 u_x + a_2(2u_x\partial_x + u_{xx}) + a_3(3u_x\partial_x^2 + 3u_{xx}\partial_x + u_{xxx})]$$
$$= [b_{0x} + i(a_1 u_x + a_2 u_{xx} + a_3 u_{xxx})]$$
$$+ [2b_0 + b_{1x} + i(2a_2 u_x + 3a_3 u_{xx})]\partial_x$$
$$+ (2b_1 + b_{2x} + i3a_3 u_x)\partial_x^2 + 2b_2\partial_x^3$$
$$= i(6u^2 u_x - u_{xxx})$$

which lead to

$$a_{3x} = 0, a_{2x} + i2ub_2 = 0, a_{1x} + i2(ub_1 + u_x b_2) = 0$$

and

$$a_{0x} + i(2ub_0 + b_1 u_x + b_2 u_{xx}) = 0$$
$$b_2 = 0, 2b_1 + b_{2x} + i3a_3 u_x = 0$$
$$2b_0 + b_{1x} + i(2a_2 u_x + 3a_3 u_{xx}) = 0, \text{ and}$$
$$b_{0x} + i(a_1 u_x + a_2 u_{xx} + a_3 u_{xxx}) = i(6u^2 u_x - u_{xxx})$$

These relations are fixed up as

$$b_2 = 0, a_3 = \text{constant} = C_0, a_{2x} = 0 \Rightarrow a_2 = 0, a_{1x} + i2ub_1 = 0$$
$$a_{0x} + i(2ub_0 + b_1 u_x) = 0, 2b_1 + i3C_0 u_x = 0$$
$$2b_0 + b_{1x} + i3C_0 u_{xx} = 0, \text{ and}$$
$$b_{0x} + i(a_1 u_x + C_0 u_{xxx}) = i(6u^2 u_x - u_{xxx})$$

which are solved to give

$$a_1 = 6u^2, \ b_1 = i6u_x, \ a_3 = C_0 = -4, \ b_0 = i3u_{xx}, \text{ and } a_0 = 6uu_x$$

Thus,

$$\vec{A} = -4 \begin{bmatrix} 1 & 0 \\ 0 & 1 \end{bmatrix} \partial_x^3 + 6 \begin{bmatrix} u^2 & iu_x \\ -iu_x & u^2 \end{bmatrix} \partial_x + \begin{bmatrix} 6uu_x & i3u_{xx} \\ -i3u_{xx} & 6uu_x \end{bmatrix}$$

$$= \begin{bmatrix} -4\partial_x^3 + 6u^2\partial_x + 6uu_x & i(6u_x\partial_x + 3u_{xx}) \\ -i(6u_x\partial_x + 3u_{xx}) & -4\partial_x^3 + 6u^2\partial_x + 6uu_x \end{bmatrix}$$

P4.7. Apply Miura transformation (3.14) to (4.10a) and (4.11a) of the KdV AKNS pair, it results to

$$[X] = \begin{bmatrix} 0 & \lambda - (U^2 + U_\eta) \\ -1 & 0 \end{bmatrix}$$

$$[T] = \begin{bmatrix} T_{11} & T_{12} \\ T_{21} & T_{22} \end{bmatrix}$$

where

$$T_{11} = -T_{22} = 2UU_\eta + U_{\eta\eta}$$

$$T_{12} = 2U_\eta^2 + 2UU_\eta + U_{\eta\eta} - 2(U^2 + U_\eta + 2\lambda)(U^2 + U_\eta - \lambda)$$

$$T_{21} = -2\left(U^2 + U_\eta + 2\lambda\right)$$

With the defined matrix Lax pair, (4.10) and (4.11) lead to the same equations $\vec{L}\psi = \lambda\psi$ and $\psi_t = \vec{A}\psi$, given in P4.5 (with ψ replaced by η and $\gamma = 0$); accordingly, the matrix AKNS equation (4.12), with the defined matrix AKNS pair, represents the mKdV equation. (It is noted that λ in T_{21} can be replaced by the equation determined by (4.10).)

P4.8. Take x-derivative of (1.26), it leads to

$$\begin{bmatrix} \Theta \\ \Psi \end{bmatrix}_{xx} = \begin{bmatrix} -i\lambda & \phi \\ -\phi^* & i\lambda \end{bmatrix}\begin{bmatrix} \Theta \\ \Psi \end{bmatrix}_x + \begin{bmatrix} 0 & \phi_x \\ -\phi_x^* & 0 \end{bmatrix}\begin{bmatrix} \Theta \\ \Psi \end{bmatrix}$$

$$= \begin{bmatrix} -i\lambda & \phi \\ -\phi^* & i\lambda \end{bmatrix}\begin{bmatrix} -i\lambda & \phi \\ -\phi^* & i\lambda \end{bmatrix}\begin{bmatrix} \Theta \\ \Psi \end{bmatrix} + \begin{bmatrix} 0 & \phi_x \\ -\phi_x^* & 0 \end{bmatrix}\begin{bmatrix} \Theta \\ \Psi \end{bmatrix}$$

$$= -(\lambda^2 + |\phi|^2)\begin{bmatrix} \Theta \\ \Psi \end{bmatrix} + \begin{bmatrix} 0 & \phi_x \\ -\phi_x^* & 0 \end{bmatrix}\begin{bmatrix} \Theta \\ \Psi \end{bmatrix}$$

If ϕ_x is a real function and $\Psi = \pm i\Theta$, then

$$\begin{bmatrix} \Theta \\ \Psi \end{bmatrix}_{xx} + (\lambda^2 + |\phi|^2)\begin{bmatrix} \Theta \\ \Psi \end{bmatrix} = \begin{bmatrix} 0 & \phi_x \\ \phi_x & 0 \end{bmatrix}\begin{bmatrix} -\Theta \\ \Psi \end{bmatrix} = \pm i\begin{bmatrix} 0 & \phi_x \\ \phi_x & 0 \end{bmatrix}\begin{bmatrix} \Psi \\ \Theta \end{bmatrix}$$

$$= \pm i\begin{bmatrix} \phi_x & 0 \\ 0 & \phi_x \end{bmatrix}\begin{bmatrix} \Theta \\ \Psi \end{bmatrix}$$

It is rewritten as

$$\begin{bmatrix} \Theta \\ \Psi \end{bmatrix}_{xx} + (\lambda^2 + |\phi|^2 \mp i\phi_x) \begin{bmatrix} \Theta \\ \Psi \end{bmatrix} = 0 \qquad (A4.1)$$

P4.9. Consider an initial value problem for (1.25a), the initial condition is a rectangular pulse $\phi(x, 0) = A[U(x + L) - U(x - L)]$, where A is a real constant. Thus,

$$\phi_x(x, 0) = A[\delta(x + L) - \delta(x - L)] = \phi_x^*(x, 0)$$

Then, (P4.1) reduces to linear Schrödinger equation (A4.1) with $\Psi = i\Theta$

$$\Theta_{xx} + \lambda^2\Theta + (|\phi|^2 - i\phi_x)\Theta = 0 \qquad (A4.2a)$$

$$\Psi_{xx} + \lambda^2\Psi + (|\phi|^2 - i\phi_x^*)\Psi = 0 \qquad (A4.2b)$$

and (1.27) leads to

$$\Theta_t = -i(2\lambda^2 - 2\lambda\phi - i\phi_x - |\phi|^2)\Theta \qquad (A4.3a)$$

$$\Psi_t = i(2\lambda^2 + 2\lambda\phi - i\phi_x^* - |\phi|^2)\Psi \qquad (A4.3b)$$

The linear Schrödinger equations (A4.2a) and (A4.2b) are written as

$$\begin{pmatrix} \Theta \\ \Psi \end{pmatrix}_{xx} + \lambda^2 \begin{pmatrix} \Theta \\ \Psi \end{pmatrix} = 0 \quad \text{for} \quad |x| > L \qquad (A4.4)$$

and

$$\begin{pmatrix} \Theta \\ \Psi \end{pmatrix}_{xx} + (\lambda^2 + A^2) \begin{pmatrix} \Theta \\ \Psi \end{pmatrix} = 0 \quad \text{for} \quad |x| < L$$

subject to jump conditions and continuity conditions at $|x| = L$

$$\begin{pmatrix} \Theta \\ \Psi \end{pmatrix}_x (|L|^+) - \begin{pmatrix} \Theta \\ \Psi \end{pmatrix}_x (|L|^-) = \mp iA \begin{pmatrix} \Theta \\ \Psi \end{pmatrix} (\pm L) \qquad (A4.5)$$

and

$$\begin{pmatrix} \Theta \\ \Psi \end{pmatrix} (|L|^+) = \begin{pmatrix} \Theta \\ \Psi \end{pmatrix} (|L|^-)$$

1. For the bound states $(\lambda^2 = -\kappa_n^2)$, the eigenfunctions Θ_n of (A4.4) are given as

$$\Theta_n = \begin{cases} \frac{x}{|x|} c_n e^{-\kappa_n |x|}, & \text{for } |x| > L \\ B_n \sin(n\pi - \lambda_n x), & \text{for } |x| < L \end{cases} \tag{A4.6}$$

where $\lambda_n = \sqrt{A^2 - \kappa_n^2}$, $n = 1, 2, \ldots, N$, N is limited by A^2. $\Psi_n = i\Theta_n$.

With the aid of (A4.6), the jump conditions and continuity conditions (A4.5) set up the following equations:

$$-\kappa_n [c_n e^{-\kappa_n L} - B_n \cos(n\pi - \lambda_n L)] = -iAc_n e^{-\kappa_n L}$$
$$= -\kappa_n \left[c_n e^{-\kappa_n L} - B_n \cos n\pi \cos \lambda_n L \right]$$

and

$$c_n e^{-\kappa_n L} = B_n \sin(n\pi - \lambda_n L) = -B_n \cos n\pi \sin \lambda_n L$$

which are combined to obtain a dispersion equation

$$\tan \lambda_n L = -\frac{\kappa_n}{\kappa_n - iA}$$

It indicates that both λ_n and κ_n are complex values. For each eigenvalue κ_n, the eigenfunction given by (A4.6) is normalized to determine the initial scattering coefficient $c_n(0)$.

2. For an unbound state, a general solution of (A4.4) for Θ is given as

$$\Theta = \begin{cases} e^{-i\lambda x} + b(\lambda)e^{i\lambda x}, & \text{for } x > L \\ a(\lambda)e^{-i\lambda x}, & \text{for } x < -L \\ Ce^{ik_0 x} + De^{-ik_0 x}, & \text{for } |x| < L \end{cases} \tag{A4.7}$$

where $k_0 = \sqrt{A + \lambda^2}$.

With the aid of (A4.7), the jump conditions and continuity conditions (A4.5) set up the following equations for the unknowns $b(\lambda), a(\lambda)$, and C and D:

$$- i \left\{ \lambda \left[e^{-i\lambda L} - b\left(\lambda\right) e^{i\lambda L} \right] + k_0 \left[C e^{ik_0 L} - D e^{-ik_0 L} \right] \right\}$$

$$= -iA \left[e^{-i\lambda L} + b\left(\lambda\right) e^{i\lambda L} \right]$$

$$e^{-i\lambda L} + b\left(\lambda\right) e^{i\lambda L} = C e^{ik_0 L} + D e^{-ik_0 L}$$

$$- i \left\{ \lambda a\left(\lambda\right) e^{i\lambda L} + k_0 \left[C e^{-ik_0 L} - D e^{ik_0 L} \right] \right\}$$

$$= iA \left[C e^{-ik_0 L} + D e^{ik_0 L} \right]$$

$$a\left(\lambda\right) e^{i\lambda L} = C e^{-ik_0 L} + D e^{ik_0 L}$$

These equations are solved for the initial reflection coefficient $b\left(\lambda; 0\right)$. The scattering data is updated via (A4.3a) and (A4.3b) to reconstruct the potentials $V_\Theta = -\left(|\phi|^2 - i\phi_x\right)$ and $V_\Psi = -\left(|\phi|^2 - i\phi_x^*\right)$, respectively. The subsequent steps are like those presented in Section 4.6.2.

Chapter 6

P6.1. Introduce the transform $s = \tanh(x)$, (4.6) becomes

$$\frac{d}{ds}(1 - s^2)\frac{d}{ds}U + \left[2 + \frac{E}{(1 - s^2)}\right] U = 0$$

Hence, $l = 1$ and $m^2 = -E = 1$; it leads to $\kappa_1 = 1$. The eigenfunction, which is proportional to the associated Legendre polynomial P_1^1, is normalized as

$$\varphi_1(x, 0) = -\frac{1}{\sqrt{2}} \operatorname{sech}(x)$$

Then,

$$c_1(0) = \lim_{x \to \infty} \varphi_1(x, 0) e^{\kappa_1 x} = \lim_{x \to \infty} \left[-\frac{1}{\sqrt{2}} \operatorname{sech}(x) e^x \right] = -\sqrt{2}$$

and

$$F(x; t) = c_1^2(0) e^{8\kappa_1^3 t} e^{-\kappa_1 x} = 2 e^{8t - x}$$

The GLM linear integral equation becomes

$$K(x, y; t) + 2e^{8t-(x+y)} + 2\int_x^\infty K(x, z; t)e^{8t-(y+z)}\, dz = 0$$

It is solved to obtain

$$K(x, x; t) = -2\frac{e^{8t-2x}}{(1 + e^{8t-2x})}$$

Substitute $K(x, x; t)$ into (4.4a), a one-soliton solution is obtained as

$$\phi(x, t) = 2\operatorname{sech}^2(x - 4t)$$

It satisfies the initial condition $\phi(x, 0) = 2\operatorname{sech}^2(x)$.

P6.2.

$$\phi(\eta, \tau) = 48\frac{3 + 4\cosh(4\eta - 64\tau) + \cosh(8\eta - 512\tau)}{[3\cosh(2\eta - 224\tau) + \cosh(6\eta - 288\tau)]^2}$$

P6.3. With the aid of the following relations

$$[X]_t = \begin{bmatrix} 0 & -\frac{1}{2}\varphi_{xt} \\ \frac{1}{2}\varphi_{xt} & 0 \end{bmatrix}$$

$$[T]_x = \frac{i}{4\lambda}\varphi_x \begin{bmatrix} -\sin\varphi & \cos\varphi \\ \cos\varphi & \sin\varphi \end{bmatrix}$$

$$[X][T] = \frac{1}{4}\begin{bmatrix} \cos\varphi & \sin\varphi \\ -\sin\varphi & \cos\varphi \end{bmatrix}$$
$$+ \frac{i}{8\lambda}\varphi_x \begin{bmatrix} -\sin\varphi & \cos\varphi \\ \cos\varphi & \sin\varphi \end{bmatrix}$$

$$[T][X] = \frac{1}{4}\begin{bmatrix} \cos\varphi & -\sin\varphi \\ \sin\varphi & \cos\varphi \end{bmatrix}$$
$$- \frac{i}{8\lambda}\varphi_x \begin{bmatrix} -\sin\varphi & \cos\varphi \\ \cos\varphi & \sin\varphi \end{bmatrix}$$

$$[X]_t - [T]_x + \{[X], [T]\} = -\frac{1}{2}\begin{bmatrix} 0 & \varphi_{xt} - \sin\varphi \\ -\varphi_{xt} + \sin\varphi & 0 \end{bmatrix} = 0$$

it implies that φ is the solution of the sine-Gordon equation $\varphi_{xt} = \sin\varphi$.

P6.4.

$$\varphi(x,t) = \pm 2i \ln \tanh \left(x - \frac{1}{4}t \right)$$

$$\varphi_{xt}(x,t) = \mp \frac{i}{2} \frac{\operatorname{sech}^2 \left(x - \frac{1}{4}t \right) \left[1 + \tanh^2 \left(x - \frac{1}{4}t \right) \right]}{\tanh^2 \left(x - \frac{1}{4}t \right)}$$

$$\sin \varphi = -\frac{i}{2} \left[e^{\mp 2 \ln \tanh \left(x - \frac{1}{4}t \right)} - e^{\pm 2 \ln \tanh \left(x - \frac{1}{4}t \right)} \right]$$

$$= -\frac{i}{2} \left[\tanh^{\mp 2} \left(x - \frac{1}{4}t \right) - \tanh^{\pm 2} \left(x - \frac{1}{4}t \right) \right]$$

$$= \mp \frac{i}{2} \left[\tanh^{-2} \left(x - \frac{1}{4}t \right) - \tanh^{2} \left(x - \frac{1}{4}t \right) \right]$$

$$= \varphi_{xt}(x,t)$$

Bibliography

Ablowitz, M. J. and P. A. Clarkson, *Solitons, Nonlinear Evolution Equations and Inverse Scattering*, London Mathematical Society Lecture Notes 149, Cambridge University Press, Cambridge, 1991.

Ablowitz, M., D. Kaup, A. Newell, and H. Segur, Method for solving the sine-Gordon equation, *Phys. Rev. Lett.*, **30**, 1262–1264, 1973.

Ablowitz, M., D. Kaup, A. Newell, and H. Segur, The inverse scattering transform - Fourier analysis for nonlinear problems, *Stud. Appl. Math.*, **53**, 249–315, 1974.

Aktosun, T., F. Demontis, and C. van der Mee, Exact solutions to the sine-Gordon equation, *J. Math. Phys.*, **51**(12), 123521, 2010.

Brauer, K., *The Korteweg-de Vries Equation: History, Exact Solutions, and Graphical Representation*, University of Osnabrück, Osnabrück, 2000.

Burgers, J. M., A mathematical model illustrating the theory of turbulence, *Adv. Appl. Mech.*, **1**, 171–199, 1948.

Colliander, J., M. Keel, G. Staffilani, H. Takaoka, and T. Tao, Almost conservation laws and global rough solutions to a nonlinear Schrödinger equation, *Math. Res. Lett.*, **9**, 659–682, 2002.

Davidson, R. C., *Methods in Nonlinear Plasma Theory*, Academic Press, New York, 1972.

Dodd, R. K., J. C. Eilbeck, J. D. Gibbon, and H. C. Morris, *Solitons and Nonlinear Wave Equations*, Academic Press, London, 1982.

Drazin, P. G. and R. S. Johnson, *Solitons: An Introduction*, Cambridge University Press, Cambridge, 1988.

Gardner, S., J. M. Greene, M. D. Kruskal, and R. M. Miura, Method for solving the Korteweg–deVries equation, *Phys. Rev. Lett.*, **19**(19), 1095–1097, 1967; doi:10.1103/PhysRevLett.19.1095.

Griffiths, G. W. and W. E. Schiesser, Linear and nonlinear waves, *Scholarpedia*, **4**(7), 4308, 2009.

Griffiths, G. W. and W. E. Schiesser, Traveling Wave Solutions of Partial Differential Equations: Numerical and Analytical Methods with Matlab and Maple, Academic Press, New York and London, 2011.

Kaufman, A. N. and L. Stenflo, Upper-hybrid solitons, *Phys. Scr.*, **11**, 269, 1975; doi:10.1088/0031-8949/11/5/005.

Kawamoto, S., An exact transformation from the Harry Dym equation to the modified KdV equation, *J. Phys. Soc. Japan*, **54**(5), 2055–2056, 1985.

Korteweg, D. J. and G. de Vries, On the change of form of long waves advancing in a rectangular channel and on a new type of long stationary waves, *Phil. Mag.*, **39**, 422–443, 1895.

Kuo, S., On the nonlinear plasma waves in the high-frequency (HF) wave heating of the ionosphere, *IEEE Trans. Plasma Sci.*, **42**(4), 1000–1005, 2014; doi:10.1109/TPS.2014.2306834.

Kuo, S., *Plasma Physics in Active Wave Ionosphere Interaction*, World Scientific, Singapore, 2018.

Kuo, S. and B. Watkins, Nonlinear upper hybrid waves generated in ionospheric HF heating experiments at HAARP, *IEEE Trans. Plasma Sci.*, **47**(12), 5334–5338, 2019; doi:10.1109/TPS.2019.2950752.

Kuo, S., *Linear and Nonlinear Wave Propagation*, World Scientific, Singapore, 2021.

Lax, P. D., Integrals of nonlinear equations of evolution and solitary waves, *Comm. Pure Appl. Math.*, **21**(5), 467–490, 1968; doi:10.1002/cpa.3160210503.

Lax, P. D., Outline of a theory of the KdV equation, in *Recent Mathematical Methods in Nonlinear Wave Propagation*, Lecture Notes in Math., vol. 1640, Springer-Verlag, Berlin and New York, 1996.

Leibovich, S. and A. R. Seebass, eds., *Nonlinear Waves*, Cornell University Press, Ithaca, and London, 1974.

Manukure, S. and T. Booker, A short overview of solitons and applications, *PDE Appl. Math.* **4**, 100140 (1–5), 2021; doi:10.1016/j.padiff.2021.100140.

Marchant, T. R. and N. F. Smyth, Soliton interaction for the extended Korteweg-de Vries equation, *IMAJ. Appl. Math.*, **56**, 157–176, 1996.

Miura, R. M., Korteweg-deVries equation and generalizations. I. A remarkable explicit nonlinear transformation, *J. Math. Phys.*, **9**, 1202–1204, 1968.

Miura, R. M., C. S. Gardener, and M. D. Kruskal, Korteweg-deVries equation and generalizations. II. Existence of conservation laws and constants of motion, *J. Math. Phys.*, **9**, 1204–1209, 1968.

Novikov, S., S. V. Manakov, L. P. Pitaevskii, and V. E. Zakharov, *Theory of Solitons*, Consultants Bureau, New York, 1984.

Robinson, P. A. Nonlinear wave collapse and strong turbulence, *Rev. Modern Phys.*, **69**(2), 507–574, 1997.

Sawada, K. and T. Kotera, A method for finding N-soliton solutions of the K. d. V. equation and K. d. V.-like equations, *Progr. Theroret. Phys.*, **54**, 1355–1367, 1974.

Schmidt, G., *Physics of High Temperature Plasmas*, 2nd Edition, Academic Press, New York, 1979.

Seadawy, A. R., Exact solutions of a two-dimensional nonlinear Schrödinger equation, *Appl. Math. Lett.*, **25**, 687–691, 2012.

Stenflo, L., Upper-hybrid wave collapse, *Phys. Rev. Lett.*, **48**(20), 1441, 1982.

Taha, T. R. and M. J. Ablowitz, Analytical and numerical aspects of certain nonlinear evolution equations. III. Numerical, Korteweg-de Vries equation, *J. Comp. Phys.*, **55**, 231–253, 1984.

Whitham, G. B., *Linear and Nonlinear Waves*, John Wiley & Sons, New York, 1999.

Zabusky, N. J. and M. D. Kruskal, Interaction of solitons in a collisionless plasma and the recurrence of initial states, *Phys. Rev. Lett.*, **15**, 240–243, 1965.

Zakharov, V. E. and A. B. Shabat, Exact theory of two-dimensional self-focusing and one-dimensional self-modulation of waves in nonlinear media, *Soviet Phys. JETP*, **34**(1), 62–69, 1972.

Zakharov, V. E., Collapse of Langmuir waves, *Zh. Eksp. Teor. Fiz.*, **62**, 1745–1759, 1972; *Soviet Phys. JETP*, **35**(5), 908–914, 1972.

Index

Printed in the United States
by Baker & Taylor Publisher Services